W0106776

TECHNOLOGY AND CONTEMPORARY LIFE

PHILOSOPHY AND TECHNOLOGY

Series Editor: PAUL T. DURBIN

Editorial Board

Albert Borgmann, *Montana*

Mario Bunge, *McGill*

Edmund F. Byrne, *Indiana –
Purdue at Indianapolis*

Stanley Carpenter, *Georgia Tech*

Robert S. Cohen, *Boston*

Ruth Schwartz Cowan, *SUNY –
Stony Brook*

Hubert L. Dreyfus, *California –
Berkeley*

Bernard L. Gendron, *Wisconsin –
Milwaukee*

Ronald Giere, *Minnesota*

Steven L. Goldman, *Lehigh*

Virginia Held, *CUNY*

Gilbert Hottois, *Université Libre
de Bruxelles*

Don Ihde, *SUNY – Stony Brook*

Melvin Kranzberg, *Georgia Tech*

Douglas MacLean, *Maryland*

Joseph Margolis, *Temple*

Robert McGinn, *Stanford*

Alex Michalos, *Guelph*

Carl Mitcham, *Polytechnic
University*

Joseph Pitt, *Virginia Polytechnic*

Friedrich Rapp, *Dortmund*

Nicholas Rescher, *Pittsburgh*

Egbert Schuurman, *Technical
University of Delft*

Kristin Shrader-Frechette,
South Florida

Elisabeth Ströker, *Cologne*

Ladislav Tondl, *Czechoslovakia*

Marx Wartofsky, *CUNY*

Caroline Whitbeck, *M.I.T.*

Langdon Winner, *R.P.I.*

Walther Ch. Zimmerli, *Technical
University Carolo-Wilhelmina,
Braunschweig*

OFFICIAL PUBLICATION OF
THE SOCIETY FOR PHILOSOPHY AND TECHNOLOGY

PHILOSOPHY AND TECHNOLOGY
VOLUME 4

TECHNOLOGY AND CONTEMPORARY LIFE

Edited by

PAUL T. DURBIN

University of Delaware

D. REIDEL PUBLISHING COMPANY

A MEMBER OF THE KLUWER ACADEMIC PUBLISHERS GROUP

DORDRECHT / BOSTON / LANCASTER / TOKYO

Library of Congress Cataloging-in-Publication Data

Technology and contemporary life / edited by Paul T. Durbin.
   p.   cm. — (Philosophy and technology; v. 4)
  "A significant portion of the papers included in this volume were
first presented at the third international meeting of the Society for
Philosophy & Technology, held at Twente University of Technology
in Enschede, the Netherlands, in 1985" — Pref.
  "Official publication of the Society for Philosophy and
Technology."
  Includes index.
  ISBN-13: 978-90-277-2571-4   e-ISBN-13: 978-94-009-3951-6
  DOI: 10.1007/978-94-009-3951-6
  1. Technology—Philosophy. 2. Technology—Social aspects.
I. Durbin, Paul T. II. Society for Philosophy & Technology (U.S.)
III. Series.
T14.T3837 1987   87–28700
306'.46—dc 19   CIP

Published by D. Reidel Publishing Company.
P.O. Box 17, 3300 AA Dordrecht, Holland.

Sold and distributed in the U.S.A. and Canada
by Kluwer Academic Publishers,
101 Philip Drive, Norwell, MA 02061, U.S.A.

In all other countries, sold and distributed
by Kluwer Academic Publishers Group,
P.O. Box 322, 3300 AH Dordrecht, Holland.

All Rights Reserved
© 1988 by D. Reidel Publishing Company, Dordrecht, Holland
and copyrightholders as specified on appropriate pages within.
Softcover reprint of the hardcover 1st edition 1988
No part of the material protected by this copyright notice may be reproduced or
utilized in any form or by any means, electronic or mechanical,
including photocopying, recording or by any information storage and
retrieval system, without written permission from the copyright owner

TABLE OF CONTENTS

PREFACE   vii

A SYMPOSIUM ON ALBERT BORGMANN'S
*TECHNOLOGY AND THE CHARACTER OF
CONTEMPORARY LIFE*   1

I. STANLEY R. CARPENTER / A Discussion   1

II. MANFRED STANLEY / A Critical Appreciation   13

III. ALBERT BORGMANN / Reply   29

ALEXANDER BARZEL / The Co-Relational Community and
Technological Culture   45

EDMUND F. BYRNE / The Labor-Saving Device: Evidence of
Responsibility?   63

SYMPOSIUM ON APPROPRIATE TECHNOLOGY   87

I. STANLEY R. CARPENTER / A Conversation Concerning
Technology: The "Appropriate" Technology Movement   87

II. THOMAS SIMON / Appropriate Technology and Inappropriate
Politics   107

DANIEL CÉRÉZUELLE / Reflections on the Autonomy of
Technology: Biotechnology, Bioethics, and Beyond   129

v

DANIEL O. DAHLSTROM / *Lebenstechnik und Essen*: Toward a Technological Ethics after Heidegger   145

LARRY HICKMAN / The Phenomenology of the Quotidian Artifact   161

SYMPOSIUM ON INFORMATION TECHNOLOGIES   177

I. PAUL LEVINSON / Impact of Personal Information Technologies on American Education, Interpersonal Relations, and Business, 1985–2010   177

II. S. MUTHUCHIDAMBARAM / Information Technology, Citizens' Rights, and Personnel Administration   193

JOSEPH MARGOLIS / History, Nature, and Technology   217

ANDRIES SARLEMIJN AND PETER A. KROES / Technological Analogies and Their Logical Nature   237

KRISTIN SHRADER-FRECHETTE / Public and Occupational Risk: The Double Standard   257

WALTHER CH. ZIMMERLI / Variety in Technology, Unity in Responsibility?   279

EDMUND F. BYRNE / Work and Technology: A Bibliographical Essay   295

INDEX OF NAMES   315

PREFACE

Nearly everyone agrees that life has changed in our technological society, whether the contrast is with earlier stages in Western culture or with non-Western cultures. "Modernization" is just one of various terms that have been applied to the process by which we have arrived at the peculiar lifestyle typical of our age; whatever the term for the process, almost all analysts agree in finding *technology* to be one of its key ingredients.

This is the judgment of critics of all sorts – anthropologists, historians, literary figures, sociologists, theologians. Volume 4 in the *Philosophy and Technology* series brings the perspectives of philosophers to bear on the issue of characterizing contemporary life, mainly in high-technology societies. Some of the philosophers look at the issue directly. Others focus on work life – or on the living arrangements that surround or condition or offer refuge from work life in technological society. Still others reflect on particular technologies, especially biotechnology and computer technology, that are increasingly affecting both work and family life. There is also a paper on the nature of thinking in technological *praxis*, along with two papers on whether it is appropriate to export this sort of thinking to Third World countries, and another paper on the issue of responsibility in technology – which would have fit better in volume 3 of the series, entitled *Technology and Responsibility* (1987). Finally, volume 4 closes with a broad-ranging bibliography that takes work and technology as its focus.

Philosophers may have been late in turning to these issues – with occasional exceptions such as Hannah Arendt. But now that they are discussing the issues with increasing frequency, we can hope for new and exciting insights. A significant portion of the papers included in this volume were first presented at the third international meeting of the Society for Philosophy & Technology, held at Twente University of Technology in Enschede, The Netherlands, in 1985. Others come from symposia in the United States run by the Society in conjunction with various meetings of the American Philosophical Association. Two papers came in "over the threshold," simply mailed in to the editor – a practice endorsed by the Society.

Paul T. Durbin (ed.), Technology and Contemporary Life, vii–viii.
© 1988 *by D. Reidel Publishing Company.*

Once again acknowledgments are due, especially words of gratitude to the secretaries in the Philosophy Department at the University of Delaware, Mary Imperatore and Dorothy Milsom, and to the referees for all the papers in the volume. The latter will remain unnamed here, but their work is warmly appreciated. Finally, thanks are due once again to the helpful and efficient staff of the D. Reidel Publishing Company.

STANLEY R. CARPENTER

A SYMPOSIUM ON ALBERT BORGMANN'S *TECHNOLOGY AND THE CHARACTER OF CONTEMPORARY LIFE:*

I. A Discussion

"For six millennia at least, the banks of the lower Nile have been a human artifact rather than the swampy African jungle which nature, apart from man, would have made it." So observes historian of technology Lynn White, Jr.[1] The myth of Prometheus equates human uniqueness with our ability, indeed our need, to fashion artifacts in order to assume a place among the flora and fauna of the planet. Our appropriation of the world is forever mediated by our tools. We swim in an artificial sea.

It may be argued, however, that during the most recent several centuries a profound shift has occurred in the relationships existing between humans and their tools. This span of time is an exceedingly small part of the history of our species, a mere blink of an eye. For those of us most directly affected, the inhabitants of the world's industrialized countries, a profound uneasiness would seem to accompany this new mode of tool use. On the one hand many of us live better than kings of old in a world of material commodities. On the other, we may sense a certain shallowness accompanying our affluence. Life appears to lack a center, an ultimate concern.

The work which we are discussing in this symposium, Albert Borgmann's *Technology and the Character of Contemporary Life*,[2] attempts to diagnose our uneasiness, and to suggest measures that can restore the center to our lives, even as technological change continues apace about us. Borgmann has written a thoughtful and profound book. It is both sweeping in its breadth and meticulous in detail. Its tone is one of grace and civility.

The focus of the book is the technology of the modern industrial era. Despite the fact that human tool use is ancient, Borgmann argues that modern technology is qualitatively unique. In a sense carefully spelled out, earlier periods are described by him as "pre-technological."

How is one to take the measure of such a complex cultural phenomenon as technology? Borgmann rejects substantive theories which cast technology in a deterministic role. The thought of Jacques Ellul comes

1

Paul T. Durbin (ed.), Technology and Contemporary Life, 1–12.
© 1988 *by D. Reidel Publishing Company.*

most readily to mind. A firm commitment to human freedom and a belief in human potential to create and recognize significance undergirds and lends a hopeful tone to Borgmann's writing. Likewise, instrumentalist positions, embodying the "neutrality of technology" concept are also rejected. Rather than diverting our eyes from technological practice toward social and political institutions, it is technology itself that he holds under the microscope for sustained scrutiny.

The modern industrial era embodies a novel mode of world appropriation – a mode shaped and demarcated by what Borgmann terms the "device paradigm." "Things" of pre-technological eras and "devices" of the present era are contrasted. With a pre-technological thing – say, a hearth – one is not only shielded from the cold but also provided with a place for family and friends to gather. The fire is sustained by the family members, and the hearth in turn repays the efforts with comfort, context, and focus. When the hearth is replaced with a central heating device, this focus of social interaction is displaced, if not lost. A commodity, warmth, is conveniently procured while the device for its procurement is rendered invisible. What is lost in the process may seem to be intangible but it is nonetheless significant.

The hearth/furnace case is an exemplar of the device paradigm. Means become separate from ends. Both acquire distinctive characteristics and a life of their own. Means possess a protean character. Heating or cooling devices, for example, take numerous forms and utilize varied fuels. An understanding of the device, usually concealed from the user, becomes restricted to experts who, one can only hope, know how to repair it when it breaks. Ends become commodities, warmth, transportation, sustenance, etc.

We may be inclined to dismiss the losses accompanying the device paradigm. In virtue of its progressive application, toil, poverty, and suffering have given way to liberation, enrichment, and conquest. At the same time, Borgmann argues, a price for such progress is to be paid. Disburdenment from adversity can at the same time lead to disengagement, distraction, shallowness. Such is the case when technology becomes for us little more than an ensemble of commodities mysteriously emerging out of a background which we neither comprehend nor control.

World appropriation in terms of the device paradigm continues to increase. As Langdon Winner has aptly put it, we are confronted with a world of "material accomplishments and social adaptations of astound-

ing completeness."[3] Reaction to this fact may take the form of resignation and passivity. What efforts are made to meliorate technological
mistakes or excesses amount to reforms *within* the milieu of technological practices themselves. While Borgmann does not ignore the beneficial
effects of "technological fixes," he is explicit in demanding more. What
is called for is a reform *of* technology, a task requiring a new consciousness, a new mode of world appropriation.

In this regard, Borgmann's thought bears the stamp of Heidegger's
views on technology. Indeed, I find Heidegger *à la* Borgmann significantly more intelligible than the great phenomenologist's own phraseology. At the same time, Borgmann not only lends perspicuity to the
discussion, but also advances and refines Heidegger's insights. This
seems particularly true with regard to the question of the reform of
technology.

Heidegger, in his later writings, abandoned the Cartesian obsession
with epistemological certitude. In particular, he denied the possibility of
formulating a culturally independent foundation for all knowledge. He
came to believe that significant knowledge was disclosed in the particular event or object. Heidegger, so Borgmann contends, began to see
technology as a force that intrudes upon our intercourse with the world
of particular and significant things. While Borgmann shares Heidegger's
diagnosis of the loss of significance accompanying the technological
mode of appropriating the world, he does not adopt Heidegger's anti-
dote. Heidegger appears to recommend an escape from the technological into pre-technological modes of world engagement. To his credit,
Borgmann rejects this tack. While supporting Heidegger's judgment as
to the stifling effects of the rule of technology, he isolates particular
events and practices that possess the potential for engaging us and
imbuing daily life with meaning and purpose. Such particulars he calls
"focal concerns."

Rather than casting technological practices and focal concerns as
opposing categories, Borgmann claims that the focal concern may
actually be *heightened* by the technological context. He is clear, however, that such need not be the case. It is only when the rule of
technology is disclosed, that is, when our commodity-based style of
living is seen for what it is that focal concerns can begin to take on
importance.

The family meal, one of Borgmann's favorite examples, potentially
embodies a focal concern. We are all too familiar with the normal case.

The modern U.S. family may, with luck, converge at about the same hour of the day to ingest precooked delights from colorful cardboard boxes prepared in some unknown factory. While satiating their hunger, the family members may find it impossible to converse owing to the press of other commitments or to the ever-present drone of the television. Kitchen clean-up involves the depositing of tinfoil trays in plastic bags and the twirl of the knobs of the dishwasher, disposal, or trash compactor.

The family meal could be different. It could represent a cooperative and loving preparation of food by all of the family members. The joys and crises of the day could be talked out around the table. Emotional bandages could be applied and the family bonds and common concerns reinforced. Of course, such will not be the case while we accept the modern lifestyle as the inevitable price of modernity.

Borgmann cites several activities with the potential of becoming focal practices, including, in addition to the culture of the table, playing music, gardening, hiking, running. Focal concerns embody the civility, decency, and gentleness that pervade Borgmann's writing throughout this entire book. For him, a focal concern isolates and restricts the technological appropriation of the world. Instead, values of world citizenship, knowledge of the world, gallantry, intelligence, valor, refinement, and charity acquire specific focal embodiment.

The task of creating focal practices within daily life is uniquely challenging to the citizens of a technological society. Like Heidegger, Borgmann sees pre-modern civilizations as having possessed more naturally the centripetal forces that lent import to art, prophecy, artisanal skills. Biological milestones such as birth, marriage, or death were endowed with special meaning. As Alasdair MacIntyre has described it,[4] practices were embodied in institutions: church, guild, manor, patriarchy. Understanding such practices amounted to what Borgmann calls "deictic explanation." "The distinctive feature of a deictic explanation is not its method, but its subject, something unique and concrete that is at the center of attention and of its world, a holy place, for instance, that focuses and orients the world around it."[5]

The protracted collapse of feudalism and the gradual emergence of market societies weakened and ultimately all but obliterated the deictic mode of defining daily living. The causes have been oft repeated: the need for money for the Crusades leading to the monetization of dues and the Enclosure Laws; the voyages of discovery and increased commerce; the growth of cities; changes in religious attitudes concerning money lending, etc.

With these events and abetted by the later emergence of industrial technology came an all-important change in human perceptions concerning role, livelihood, and destiny. The deictic mode of embracing life was progressively replaced with the device paradigm. Land, labor, and capital as commodities were literally created. As structured by the device paradigm, social and political life suffered a diremption into means on the one hand and ends on the other, machineries versus commodities. The result of this split was the establishment of the "commodity basis of social life."

Borgmann discusses in detail the concomitant emergence of liberal democratic thought, which aimed to foster formally just and economically fair societies but which stopped short of defining the good society teleologically.

One can understand the reluctance of liberal political theorists to prescribe the good life considering the sectarian past against which they were reacting. Their vision, however, was more than incomplete, Borgmann contends. We are given an important clue here to the central role which he affords to the device paradigm. Whereas liberal democracy leaves open the nature of the good society, its realization, Borgmann argues, is not left open but is instead accomplished technologically. Ends are procured by means of technology, and take the form of commodities. In his words, "liberal democracy is enacted as technology. It does not leave the question of the good life open but answers it along technological lines."[6]

Nowhere is this more apparent than in the case of work. Borgmann wisely avoids romanticizing pre-technological work. At the same time he shows that the commodity basis of social life is enacted, defined, and constrained along lines dictated by the device paradigm. Division of labor, with its lessened dependence upon skill and dexterity, is promoted as demonstrably more productive. Divided labor is tied to the machine and stripped of its potential as a focal human practice, becoming instead one more mechanical component. E. F. Schumacher has observed that this degradation of work transforms it into a "necessary evil." He writes:

From the point of view of the employer, it is . . . simply an item of cost, to be reduced to a minimum. . . . From the point of view of the workman, it is a "disutility"; to work is to make a sacrifice of one's leisure and comfort.[7]

Borgmann notes the special irony accompanying the degradation of work. Particularly in the United States it is work that is regarded as bestowing dignity upon the individual. By contrast, receipt of welfare payments earns our contempt. Borgmann is not sanguine in his assessment of proposed reforms of work through automation or in terms of the reorganization of labor relations. He observes:

> The framework of work is constrained by the intricacy of the productive machinery in the widest sense. . . . Crucial and growing parts of the technological machinery are becoming intellectually and physically ever more remote from the common person's competence. The circle of expertise is suffering a final contraction and centralization.[8]

Without significant reforms, he believes, the culmination of this tendency will be the elimination of work for the majority.

How can technology be reformed? As already noted, the character of contemporary life is defined by the device paradigm. To the extent that there is political consensus that the promise of liberal democratic values is always in need of fuller realization, the technological society *does* self-correct. Our air is purer today thanks to environmental legislation. Automobiles are markedly more fuel efficient. Borgmann offers detailed and sustained argument, however, that most calls for reform fail to get at the root causes of social dysfunction. In this regard he discusses and dismisses all of the following: the microelectronic transformation of work; the beautification of technology through increased attention to its aesthetic potential; increased public discussion of human values in the context of technology; the democratization of technology and a commitment to worldwide distribution of technological culture.

Time does not permit our doing justice to each of these topics. Borgmann does not minimize the potential good that such reforms might achieve. He is insistent, however, that they do not go far enough, or, better stated, deeply enough. Each reform is a reform *within* technology and not a reform *of* technology. Any reform that does not alter our mode of appropriating the world leaves intact the device paradigm. By persisting in the belief that the *telos* of technology is consumption, we fail to subject that appropriation to moral scrutiny. By approaching the reform of technology within the device paradigm, we are like Br'er Rabbit and the tarbaby. With each lunge away from some technological excess, we sink deeper into the grip of the technological fix. We shake our heads in disbelief at the attempt to escape the terror of offensive nuclear missiles by means of a defensive space shield of unprecedented technological complexity.

Borgmann is unequivocal in the assertion that the reform of technology requires the restoration of focal concerns to the center of life. Accompanying this is the restriction of technology, both the background of machinery and the foreground of commodities, to a subsidiary role. Focal concerns should once again provide a center of meaning for individual and group alike.

Borgmann avoids the prescribing of focal concerns. Indeed, to do so would be to contradict his repeated contention that focal concerns arise out of deictic discourse. He does characterize deictic discourse by its enthusiasm, sympathy, and tolerance. As to the content of a focal or deictic concern, there is an *ad hoc* character to it. As he says, "One can never procure it or control it or build it up from small and incontrovertible pieces."[9] On the contrary, one must assume it in the sense of accepting it and responding to it.

By his choice of examples, we are given clues as to Borgmann's own candidates for personal focal concerns. As mentioned, he finds importance and meaning in the culture of the table, in learning to play a musical instrument, in running.

As to the reform of technology in a corporate sense, we are offered a number of specifics. In the first place, the *sine qua non* of public deictic discourse is a basic tolerance of others, a tolerance that can only be assured within the framework of democratic institutions. Beyond this, he cites approvingly Galbraith's assertion that "the emancipation of belief is the most formidable of the tasks of reform and the one on which all else depends."[10] For him, this translates into our freeing ourselves from the belief that technology, or life characterized in terms of the device paradigm, must be definitive. Specifically, for example, we must reject the equating of quality of life with standard of living. A growing number of individuals, we are told, are practicing "voluntary simplicity" (V.S.), according to which lower consumption levels are achieved by means of goods that are durable or reparable. The full range of V.S. practices embodies the principle that wealth does not equal affluence.

Additionally, emancipation of beliefs about technology will include a rejection of the idea that work must be so constituted as to be boring or merely tolerated as a means to a separate end. To configure work in the manner of a focal concern means attending to the manner in which tools are used: does the tool aid the worker, providing him with greater freedom of choice as to pace or quantity of output or, on the other hand, does it set the pace of work itself? Is the tool something with which the worker's skills are enhanced, or does it employ the worker as a minor and mentally stultified component of the production process? Is the production process at all related to the natural setting and to locally available materials, skills, and traditions? Is the scale of work small where possible and conducive to worker teams? In my opinion, Borgmann's ideas on focal work represent a systematic improvement of E. F. Schumacher's own writings on the subject of good work.

Borgmann's blueprint for the reform of technology contains interesting specific suggestions. He is realistic about the central and growing role that large corporations play in our society. One writer has recently predicted that the continuation of present trends would mean that by 1988 three hundred giant firms will produce half of the world's goods and services.[11]

Borgmann speculates that a public deictic discourse on the economy might take the form of a "two-sector" economy. Transportation, utilities, communications, along with machine tools, cars, and appliances would continue to be produced at their currently massive scale. Resource extraction, insurance, and finance would probably also remain large scale. On the other hand, a host of activities could be locally constituted along the lines of focal work. Included are activities pertaining to food, furniture, clothing, health care, education, and instruction in music, arts, and sport.[12]

The adoption of local, small scale, and labor-intensive production units could only come as a result of public deictic discourse. Debates would need to thrash out matters of local ownership and to devise agreed-upon measures for labor intensity, firm size, energy consumption, and acceptable capital cost per workplace. Such discourse, however, is not likely to occur as long as the patterns of daily life and the goals we pursue are interpreted solely within the device paradigm. To create the conditions for such a discourse by demystifying the encompassing role played by technology is, as I see it, Borgmann's overriding concern.

It is time to mention a few points with which I find some difficulty in Borgmann's approach. It is clear to me that there is very little that could be offered in the way of factual amendment or correction. Borgmann's mastery of historical as well as contemporary material is both comprehensive and deep. It may thus be that the issues which I raise reduce to questions of emphasis.

I share in the judgment that with the modern era the very fabric of life was profoundly altered. Additionally, it is equally clear that technology has been centrally implicated in this change. Is there a chance of misunderstanding, however, in characterizing former times as "pre-technological"? Indeed, as I noted at the outset, the myth of Prometheus seems to embody, at the deepest level, the insight that human appropriation of the world has always been technological. Lynn White, Jr., has made a career of pointing out the technological richness of the medieval era. His disagreement with Henry Adams's characterization of the Middle Ages as the age of spirituality and the modern era as the age of technology is well known.[13] White's rebuttal emphasizes both the medieval sanctification of technology embodied in the gothic cathedrals as well as the realization of more broadly drawn spiritual values such as a reduction of human slavery, which non-human power sources facilitated.

It could thus be argued that technology has possessed ontological depth for nearly the entire life of our species. When, however, we come to the modern era, it is clear that the role of technology has undergone radical transformation, the full scope of which Borgmann has eloquently described. What was it that caused the restructuring of social and political life into machinery and commodities? That technology was centrally implicated in the creation of this cleavage in the fabric of culture is well established. But one might still relegate technology to an instrumental role. As Borgmann notes, the cathedral, a magnificent technological achievement, nevertheless possessed the capacity to embody focal significance in terms of comprehensiveness, unity, accessibility, and enactment. Additionally, he describes how a reform of technology that relegates both machinery and commodity to the status of means can once again lead to the realization of focal practices without a corresponding abandonment of technological skills.

Yet, if one portrays technology in an instrumental role, as I am doing, while granting a uniqueness to its structure in modern times, the source of this novelty needs to be located. This I would do by connecting the commodity basis of modern life directly with liberal democratic theory, as Borgmann does, but with greater emphasis upon the monetization of social life and its subsequent systematic formulation in market economics. Heilbroner refers to this development as the emergence of an entirely new social institution.[14] Within this institution what emerges is nothing less than a new form of life – life established on a commodity basis. Human beings are buyers and sellers. Adam Smith places within the human mind itself the disposition to truck, barter, and exchange.[15]

The coalescing of market economics and liberal political thought produced a chain reaction of historical developments. While, as Borgmann has observed, liberal democratic theory embraced values of liberty, equality, and self-realization, it also made possible the radical diremption of social life into two realms, one of business and commerce, the other of civil concerns. The business culture was thus freed from the goals of social justice while at the same time being preoccupied with the realization of maximum individual freedom. While the civic culture might strive to realize ideals of altruism and community, the business culture was propelled by totally different forces. Two conversations were set in motion, two paradigms, if you will. Cogency in argument from one perspective was irrelevance from the other. Heilbroner writes:

It makes no more sense for economists to converse about which of two equally profitable enterprises is "better" than for scientists to ask which of two experiments is more pleasing to the gods.[16]

It is thus no mystery how Adam Smith could describe in exquisite detail the division of labor without stopping to reflect upon its possible stifling of the human spirit. By means of the emerging power technologies of the Industrial Revolution and the accessibility of displaced labor, the business culture experienced virtually unimpeded growth and power. It is thus no wonder that its use of technology has been ruthless and destructive of both human capital and the non-human world. These concerns were simply not a part of the commodity basis of life. Such a basis, Borgmann notes, tears into and gradually removes the fabric of social institutions.

At the same time, the civil culture has had to deal with the damage created by technologies set loose without moral constraints. Welfare economists euphemistically refer to such matters as "externalities." It is further the case that technologies operative within the business culture can literally preempt social and political choices that the civic culture should be given. Winner has convincingly made this point in his well known article, "Do Artifacts Have Politics?"[17]

I of course do not mean to suggest that it is only the use of technology by the business culture that needs reform. Government, too, denies its heritage when questions of global security are reduced to matters of weapons parity. It should also be added that the concept of two cultures may very well be more a matter of rhetoric than of reality.

What I would observe is that Borgmann's contention that social and political reforms of technology are not up to the task and that a reform of technology along different lines is possible leaves me somewhat puzzled. If, as he says, we need to restore the elements of the device paradigm, machinery, and commodity to their proper position as means, he would appear to be agreeing with my persistence in maintaining that technology is mainly a genuine protean instrumentality. I would see the reform of technology as dependent upon a social philosophy that is other than commodity-based or one that at least augments commodity concerns with focal ones. Borgmann's brilliant and sensitive exposition of focal practices is suggestive as to the tacks that social philosophers might pursue.

I am possibly more optimistic than Borgmann concerning the transformational potentials of steady-state economics, alternative technologies, and voluntary simplicity, that is, alternative practices within existing social contexts. I agree entirely, however, that none of these approaches makes sense except as a background to focal concerns.

In conclusion I would note that Borgmann's work allows us to confront the dailiness of life with heightened senses. The potential for focal practices exists, as he makes clear, even as we swim like fish in a sea of technological artifacts. It is Borgmann's gift to have made visible the water.

Georgia Institute of Technology

NOTES

[1] Lynn White, Jr., 'The Historical Roots of Our Ecologic Crisis', in his *Dynamo and Virgin Reconsidered: Essays in the Dynamism of Western Culture* (Cambridge, Mass.: MIT Press, 1968), p. 76.
[2] Albert Borgmann, *Technology and the Character of Contemporary Life: A Philosophical Inquiry* (Chicago: University of Chicago Press, 1984).
[3] Langdon Winner, 'The Political Philosophy of Alternative Technology: Historical Roots and Present Prospects', in D. Lovekin and D. Verene, eds., *Essays in Humanity and Technology* (Dixon, Ill.: Sauk Valley College, 1977), p. 131.
[4] Alasdair MacIntyre, 'Objectivity in Morality and Objectivity in Science', in H. Engelhardt and D. Callahan, eds., *Morals, Science, and Society* (Hastings-on-Hudson, N.Y.: The Hastings Center, 1978), pp. 28–29.
[5] Borgmann, *op. cit.*, p. 72.
[6] *Ibid.*, p. 90.
[7] E. F. Schumacher, *Small Is Beautiful: Economics as if People Mattered* (New York: Harper & Row, 1973), p. 51.
[8] Borgmann, *op. cit.*, p. 123.

[9] *Ibid.*, p. 241.

[10] *Ibid.*, p. 239.

[11] Robert B. Reich, *The Next American Frontier* (New York: Penguin, 1983), p. 266.

[12] Borgmann, *op. cit.*, p. 241.

[13] Lynn White, Jr., 'Dynamo and Virgin Reconsidered', in his *Dynamo and Virgin Reconsidered* (see note 1, above), pp. 47–73.

[14] Robert L. Heilbroner, *The Making of Economic Society* (5th ed.; Englewood Cliffs, N.J.: Prentice-Hall, 1975), pp. 47–71; see also Karl Polanyi, *The Great Transformation* (New York: Rinehart, 1944).

[15] Robert L. Heilbroner, review of Donald N. McCloskey, *The Rhetoric of Economics*, *The New York Review of Books*, April 24, 1986, p. 46.

[16] *Ibid.*, p. 47.

[17] Langdon Winner, 'Do Artifacts Have Politics?', *Daedalus* **109** (Winter 1980): 121–136.

SYMPOSIUM ON ALBERT BORGMANN
II. A Critical Appreciation

I hope you will not mind my phrasing these remarks in the form of a letter. You and I have become acquainted through correspondence. Now that I am to have the pleasure of meeting you at last and hearing your reply in person, I find it difficult reverting to the impersonal style of a formal paper. Instead I would like to celebrate our converging histories of "focal" inquiry, pursued in books and letters, by sustaining the personal idiom of address to which I have become happily accustomed.

I must begin with simple congratulations. Rigorous and demanding as it is, your book pays tribute to the sensibilities and wisdom of its author. It appeals to logic, but its standard of intelligibility requires as well the experiential history and resources of the reader. It forces one to engage with you as in a conversation, and this letter is late precisely because a final proper reading for this occasion demanded a more leisurely pace of reflection than I had anticipated. You will find criticisms and disagreements in the remarks to follow, but they concern details of a project for which I have not only deep respect but affection as well; affection because your book is among those which remind us that the life of the mind is an embodied practice, an inquiry into the conditions and possibilities of existence as sentient creatures. To disagree with you as well as to learn from you are to put one's own "ultimate concerns" into question. Your methodological care invites us to examine our reactions without narcissistic indulgence while yet allowing us to contradict you with evidence from our own lived experiences. So, gratefully indebted for your part in the moral exploration of our age, I hope these reflections you have inspired will be acceptable contributions to that larger conversation in which we are both privileged to participate.

I would like to begin with a summary of what I understand your book to be about as an integral project. Perhaps some of my later criticisms with which you may disagree will turn out to be misreadings of your intentions. Subsequent to that account, I shall turn to a number of topics on which I was not able to accommodate your views with ease. These include your treatment of Marxism; your effort to enlist "focal things and practices" in the political search for a common good (expressed

13

Paul T. Durbin (ed.), Technology and Contemporary Life, 13–28.
© 1988 by D. Reidel Publishing Company.

otherwise as your effort to set limits both to the chaotic possibilities of liberal pluralism and to my scenario of historical motion toward the "libertarian technicist society"); and your complete lack of attention to the problem of evil. Please allow me to stress the tentative and conversational quality of what is to come. Doing justice to a text as subtle as yours, much less to your intentions, must be attempted in a most provisional spirit. No one is more aware of that than I who have also risked the patience of readers on this elusive topic of the culture of technology.

I

Your project as I see it.

This account of your project is based on my encounter with your text. Your intentions must be inferred, and I take the liberty of considering aspects of the text which may not have been intended consciously. I understand you to have aspired to at least five broad accomplishments.

1. A critique of modern technology which, because phrased in non-primitivist terms, allows for the assimilation of technology into a world redeemed from its oppressive role.
2. A theory of why people observably conform to the rule of technology even though they show signs of comprehending its disbenefits.
3. A social philosophy that sets limits to liberal pluralism understood as random subjective preferences relative to a field of technological commodities. In place of that you want to ground a rational and finite pluralism in some objective features of experience as undergone by human beings in ways broadly explored in the thought of Martin Heidegger. (You wish both to assimilate Heidegger's perspective and to transcend it by eliminating his nostalgic pre-modernism and elaborating his insights from a passive stance of "waiting" into an active philosophy of practice.) On behalf of this cause you develop a philosophy of focal things based on a subtle combination of materialism and intuitionism.
4. A political philosophy that moves beyond liberalism's concern for distributive justice to a renewed classical quest for the good society. Toward this end you develop some claims about categories of intersubjectivity generated by experiences with focal things. These

categories (enthusiasm, sympathy, tolerance) you seem to regard both as a possible basis for mutual intelligibility between partisans of diverse focal experiences and as the ground for collective comprehension of a standard of political good translatable into public policies.
5. A philosophy of pedagogy that assimilates affect into a non-Cartesian understanding of cognition. I find this intention present throughout your book as you explore the relationships between focal practices and the concept of discipline. In all these discussions you persistently attempt to transcend dualisms of mind and body, subject and object.

I would like now to note where I believe your efforts to be most successful. As you know from my utilization of an earlier account of it in my own work, I find your device paradigm a major contribution to the philosophy of technology. Its originality lies not in its unfamiliarity but, to the contrary, in its organization of a host of familiar experiences and allusions. You have helped us to see a common phenomenology across a diverse spectrum of topics from television to politics. With it, you have also enriched our understanding of the concepts of "commodity" and "practice," as well as their relationships to each other, especially as regards the way a given commodity, as output of a device, degrades our practices pertaining to the experiences which that commodity once mediated before it was transformed by a device. Further, you have improved upon or opened new frontiers of social criticism by means of the device paradigm. You have used it to deconstruct common terms such as availability, engagement, freedom of choice, and opportunity. With its aid you have also forced us critically to question the assumptions of "value talk," the benefits of rearranging workplace conditions, the role of television in public life, the quality of our discourse about experience both within and outside the device paradigm, and how we choose criteria for public policy (especially pertaining to wilderness and cities). In doing all this, you made ample use of Heidegger's thought without inflicting upon us much of its obscurity.

A second feature of your book that greatly appeals to me is your effort to limit the promiscuity in our usages of the term "pluralism." Despite misgivings which I will get to shortly, I agree wholeheartedly with your insistence that pluralism must refer to something deeper than "frivolous" commodity preferences or nominal group identifications. It should be grounded, you imply throughout your book, in profoundly articulated phenomenologies of focal things and practices. The elaboration of this suggestion, I think, would constitute a contribution to pluralist theory and go a long way toward rescuing it from its present abstractions. That said, there is a tragic dimension to your argument which I think you overlook in your desire to build bridges between the diverse partisans of focal things. More of that later.

Finally, I appreciate very much what I have referred to as your contributions to the philosophy of pedagogy. Under this category I include all your efforts to explain in terms of each other the concepts of significance, depth, orientation, engagement, maturity and discipline, and to contrast these with what you mean by "frivolous" and "shallow." Your debt to Heidegger is clear. But your efforts to disengage his thought from nostalgia and passivity, and to link it to concepts of agency and practice in ways congenial with pragmatist thinking constitutes a contribution in its own right. The occasion does not permit reproducing even a few of the fruitful associations your pedagogical passages stimulated in this reader. Let me restrict myself to just one.

I refer to a segment of your most pedagogic of all chapters (number 22), entitled "The Challenge of Nature." There you speak of the inversion of nature and culture (p. 194). This beautiful passage led me on a search for further analogues to the wilderness. It occurred to me that one was the Past itself, the wilderness of "all that happened" – not a calm place to be sure, but a place increasingly domesticated by the secular de-mystifications of technical historiographical mapping and social scientific explanation. As our pre-technological ancestors were once prisoners of the wild compared with the control technology now affords us over nature, so – you point out – our childhood will was circumscribed by the authority of our parents now frail and old as we stand liberated by adulthood. So too, I would add, our Past, safely disenchanted by the Enlightenment of superbeings and sacred claims, is dead and gone leaving us free on new frontiers. Is History, then, "bunk" as Henry Ford and most current undergraduates would have it? What is the "new respect" we owe it? Is it worthy only to provide for our amusement the gameparks of memory? As your parental analogy conducts us to the philosophic center of the Welfare State, so does my analogy reveal an instance of why we find such agonizing ambiguity in the concept of education.

Many such analogies, associations, and reflections were stimulated by your text, but it is difficult to praise your work piecemeal. Part of its inspiration derives from its synoptic ambition, its sweep and scope of effort. Therein lie some of its problems as well, to which it is now time to turn. I shall enclose my criticisms under two broad headings: your treatment of Marxism, and of the relationships between good and evil.

II

Marxism:

Why did you make such a strenuous effort to dissociate your thought from that of Marxism? I made the same effort, not even engaging as much with Marxism as you did. I was persuasively criticized for that, sufficient to make me examine my motives for having omitted the connection. There seemed to be two main reasons. One was the desire to avoid being ideologically labeled. A second was an unwillingness to submerge myself in yet another massive and complex literature sufficiently to do it justice. I now regard that to have been a mistake. The consequence was a book tilted more than I wanted toward an a-sociological idealism, almost to the point of linguistic determinism. As you do, I had much of sociological import to say, but managed to do so with almost no reference to social structure and organization. I think it is an intellectual fact of our time that thinkers inspired by the Marxist project in general produce the most sophisticated sociological and historical accounts of how social structure mediates things and ideas in our century. They are the ones who most successfully discuss matters without, as you say, "raising the question of values." Even those of us who do not think of ourselves as ideologically Marxist cannot afford to neglect the cumulative treasury of this work.

You give what I take to be two reasons for downplaying this literature, one with which I partially agree, and one which I think is misguided. The first reason is that Marxists (even Marx) hold technology *per se* to be morally neutral, discounting its influence in favor of the social relations of production. I agree with this in the sense that one does not find in almost any Marxist analyses the direct confrontation with physical technology and its phenomenologies which you urge upon us. On the other hand, in your effort (*via* your comment on Marcuse, p. 84) to reduce Marxists to a combination of substantivist and instrumental theorists on technology (both of which you criticize in chapter two), you underrate the subtlety of what they mean by the forces of production. One of the closest readings of *Capital* is that of David Harvey (*The*

Limits to Capital) who is also in virtually complete command of the Marxist corpus. He reminds us that:

> The identification of "technology" with the "forces of production" is erroneous and the mainspring of that misreading of Marx that turns him into a technological determinist. Technology is the material form of the labor process through which the underlying forces and relations of production are expressed. To equate technology with productive forces would be like equating money, the material form of value, with value itself, or equating concrete with abstract labor. But in the same way that an analysis of money can reveal much about the nature of value, so an analysis of actual technologies can "disclose" the nature of the productive forces and the social relations embedded within the capitalist mode of production (p. 100).

This passage seems to me to escape classification as either substantivist or instrumentalist in the way you seem to mean these.

Your second stated reason for downplaying Marxism, so far as I can tell, is your impression that Marxists believe "the essential choices that most people make are directed not by some concealed and consequential pattern but by other people who, at least in the Western democracies, constitute a small and definite minority class" (p. 82). Interestingly you distinguish between vulgar Marxism (the "crude thesis") and more sophisticated forms. Yet your view of orthodox Marxism gets illustrated and developed on the crude level at which "exploitation is defined as the maximizing of profits by the corporations at the expense of the consumer" (p. 82). Among the ways in which crude Marxism is commonly separated from higher versions, two in particular stand out as important for assessing your discussion. One is that high Marxism is never presented as any form of conspiracy theory. The other is that high Marxism never reduces the concept of exploitation to the level of moral psychology. High Marxism is always about the historical sociology of knowledge, the interplay between structure and consciousness. With the exception of a few words on Habermas, Sweezy, and Braverman, your consideration of Marxism remains at the level of what you call the crude thesis. This is in striking contrast with your treatment of liberal thinkers like Rawls and Dworkin. Two illustrations are what you say about wealth and about Marxism's "primal and final error."

Regarding wealth, you grant that the rich are disproportionately well represented in politics, and that they have the power to protect their privileges. "But," you conclude, ". . . those privileges are in the main either vacuous or exercised in consonance with popular goals." You finally reduce the issue of wealth to one of ethics: extravagant consumption. "But," you say, "if the wealth, so consumed, were to be distributed over the entire population, the general benefit would be small" (p. 83). A higher Marxism never reduces the issue of wealth to the habits of "the rich." Rather, wealth has to do with how a society accumulates and disposes of capital, and how these institutions of accumulation and disposition create opportunities and constraints relative to powers of decision in matters of production, employment, product lines, uses of technology, and political agenda setting.

With respect to the "primal and final error of Marxism," you seem to define that as "blame for the problems of the technological society . . . attributed to the selfish interests of a determinate class . . ." (p. 84). Higher Marxism never moralizes about selfishness, just as it does not speculate about conspiracies. Indeed, it sometimes acknowledges feelings of powerlessness and determinism among capitalists, as does Marx when discussing how capitalists experience crises. (Even in their own "vulgar" moments, as in the *Communist Manifesto*, Marx and Engels pay tribute to the noble motives and accomplishments of capitalism.) What high Marxism does is to depict how reality appears to the participants on the "surface" of social worlds as a consequence of the deep structures of organized production through which people suffer, survive, or prosper economically. High Marxism is on the frontiers of debate not about the motives of capitalists, but about the meaning and sources of exploitation (John Roemer, Robert Paul Wolff), the nature of capitalist crises (David Harvey, Jürgen Habermas, William Connolly, James O'Connor), the relative importance of economic anthropology (Maurice Godelier), the meanings of economic and social "structure" (Louis Althusser), the role of historiography in showing how populations came to fight or acquiesce to the contemporary structures of power in capitalist societies (E. P. Thompson, David Noble, Herbert Gutman, Eugene Genovese), and so on.

It is ironic that, in avoiding these considerations, you nonetheless sometimes adopt forms of argument which you reject Marxists for using. For example, noting the possibilities for the reform of capitalism which people do not take advantage of, you say that, "Marxists must resort to a massive exculpation of the people to save their case. They must explain popular passivity by reference to the power of advertisement, the threat of unemployment and police brutality, the promise of a high standard of living, the subversion of the mass media, and other more subtle modifications of the climate of opinion and action. But if all these factors were tools of domination, one should also discover evidence of resistance in regard to each if not revolt" (p. 84). Aside from the fact that there is ample evidence of such resistance, and of power, both subtle and forcible, used to overcome it (for a general review, see Michael Best and William Connolly, *The Politicized Economy*), you yourself resort to similar "exculpations." You ask, for example, "With what kind of consciousness are such decisions [to confirm or protest the rule of technology] made?" In response you present us with a paragraph describing what you mean by a relation to technology that is "neither one of domination by technology nor one of conscious direction of technology" but rather "one of living in technology" (pp. 104–105). Am I wrong to be strongly reminded by this paragraph of Marx's methodology in his chapters on the mystery of commodities in volume one of *Capital*, and also of Georg Lukacs's method of analyzing reification and the consciousness of the proletariat in his *History and Class Consciousness*? Your exculpatory analysis of persistent inequality on pp. 111–112, while devoted to explaining the stability of the reign of technology, also strikes me as a rather reasonable account of the dynamics of false consciousness. Indeed, what would be your reaction to the general suggestion that your whole perspective on engagement with focal things and practices, and the dynamics of enthusiasm, sympathy, and tolerance, is not far removed from what Marx seemed to mean by unalienated existence in his *Economic and Philosophical Manuscripts of 1844*?

Finally, by ignoring analyses of capitalism, I believe you deprive yourself of the right to conclude, as you do, that, "When modern men and women today make their decisions as voters or consumers, technology, as a rule, has provided the fundaments of their choices" (p. 103). When you give examples of people's choices, like going to eat at fast food places instead of cooking at home, replacing rather than learning to repair something, turning to the distractions of television instead of the leisured engagement with focal things, or moving from city to city instead of living a rooted kind of life, you speak of substantivist technological variables such as ease of procurement and availability, addictive powers, and the like. You say nothing of factors having to do with the institutional organization of capitalist work and power as experienced in such things as the affordability of commodities, the frequent need for two jobs or two income families to "make ends meet," the compression and often disappearance of free time within capitalist labor, the destruction of community economies through unregulated plant closings, the unpredictability of layoffs, the vicissitudes of economic security through business cycles, etc.

I do not want to be misunderstood. None of this is to downgrade the authority and importance of what you have to say about technology, and your conclusion that "we can make room for the discussion of the good life only if we have before us the pattern of technology and the peculiar way in which, under the semblance of a mere tool, it preforms our discourse and agenda" (p. 101). As in the case of my own work too, however, by ignoring the best of high Marxism (under which I include much of Max Weber), I think we missed the opportunity to integrate our thought with a particularly powerful form of sociology. Thus I risked idealism by seeming to suggest an autonomous power of language, and you, I think, risked materialism by stressing (despite your disclaimer) a substantivist technological determinism beyond the limits of your intention. I think both our arguments would have better served as important corrections to high Marxist sociology rather than as alternatives to it. It would be better all around, perhaps, if that corpus of work could simply be called the sociology of capitalism, leaving us free to make of it what we will ideologically since some of it converges with propositions that emanate as easily from the ideological center or even right (e.g., the cooptation of regulative agencies, the functions of the "new class," etc.). It is time to turn to the next topic.

III

Good and evil:

You appear to invest considerable faith in deictic discourse about focal things and practices as a source of consensus on a common good, sufficient even to transcend liberalism's more limited understanding of the just society. You do this in several places, but allow me to consider in detail one of them, pp. 176–179.

You begin by reminding us that "The critique of technology has two necessary and jointly sufficient conditions. The first is the existence of a concern of ultimate significance that one sees threatened by technology. The second is a profound regard for one's fellows" (p. 176). Taking off from the supposition that "if the welfare of humankind truly concerns me, then I will want to join the two concerns [my ultimate concern and human welfare] and act on that joint concern to ensure its welfare," you proceed to derive "enthusiasm, sympathy, and tolerance" as attitudinal consequences of deictic discourse. You link all this to democracy, claiming that "a revival of democracy requires the restoration of deictic discourse" (p. 177). I read you as strongly implying that deictic discourse will also help generate democracy *via* the three virtues cited above. You go on to distinguish between the simple liberal tolerance of technological democracy and tolerance that springs from enthusiasm and sympathy. Won't the latter prove oppressive or stalemate the political arena? It isn't any more oppressive than technological tolerance, you reply, because under the rule of the latter, people in any case "have no real opportunity to become ranchers, weavers, wheelrights, poets or musicians. It is increasingly difficult for them to be faithful to a place and to persons. And the technological society often does leave them alone, i.e., isolated or institutionalized, when that is anything but desired" (p. 178). That condition, I would suggest to you, is due not to technology but to capitalism – its labor requirements and its need both to predict the rates and types of commodities produced by technology and to produce commodities fit for rapid circulation through the markets of society.

Your longer range claim is that deictic political discourse, based on a deeper understanding of ends than Rawls provides with his notion of dominant ends, "is most likely to create conditions of collective assent and the basis of common action" (p. 179). Cashing in on this general argument you lead us along many interesting highways and byways of thought. On the whole, however, I find the analysis unconvincing. In joining argument with you, I will try to take with the seriousness you request your understanding of focal things as generating ends of "unsurpassable depth" (p. 218). My criticisms will take two interrelated forms. First, I question your implied theory of intersubjectivity which invites us to believe in your solution to the problems of the common good. I will try to show instead that a profound pluralism of inclusive ends exists whose discovery is somewhat obscured by how you choose to illustrate focal things and practices. Second, I will try to relate this point to an observation about your notion of intersubjectivity: its total neglect of the possibility of dark ends. In short, your analysis does not solve the problem of pluralism but deepens it to a level appropriate to a world of focal things, and, second, you have ignored the problem of evil.

Pluralism:

What is the point of engagement with focal things and practices? By accepting to a considerable extent the Weberian dramaturgy of modernization as world disenchantment, you more or less invite us to accept engagement with focal things as self-justificatory; a kind of "these things are what life is all about" argument. You provide some wonderful examples of what it can mean to say this, but leave things obscure enough to imply that all focal things somehow form a common kingdom of redemptive ends, ends capable of inducting us into a commonwealth of enthusiasm, sympathy, and tolerance. This is illustrated by your quotation from George Sheehan (pp. 212–213) on the appreciation of diversity. You do not deny the plurality of focal concerns but seek to "discern a unity underlying the plurality" (p. 213).

I would like to posit a simple classification of ultimate concerns that points up their diversity beyond the limits of your discussion, and yet remains true to your anti-subjectivist fidelity to experience (p. 202). This typology is based on the assumption that all of our concern with focal things is rooted in projects of self-transcendence, i.e., ways of discovering our contexts, which is what I understand you to be talking about throughout this book. I will try to distinguish between senses of self-transcendence as well as the ends appropriate to these senses.

One sense of self-transcendence we may call *Contemplation*. The point of contemplation, its "end," is revelation; allowing things to shine forth as they are to us. The receptive stance of the mystic, the sage, or the aesthete is appropriate to this mode of self-transcendence. Insofar as, "It is the dignity and greatness of a thing in its own right that give substance and guiding force to the notion of complexity," computers can be focal things, as you recognize (p. 217). But you add that "the focal significance of work with computers seems precarious." When recognizing contemplation as a form of self-transcendence in its own right, the issue of work does not yet arise. It is surely the contemplative capacity of the human spirit which accounts for the appeals of formalism celebrated by structuralists like Claude Lévi-Strauss and memorably illustrated in his great "confession," *Tristes Tropiques*, with its classic closing lines:

Farewell to savages, then, farewell to journeying! And instead, during the brief intervals in which humanity can bear to interrupt its hive-like labours, let us grasp the essence that may be vouchsafed to us in a mineral more beautiful than any work of Man; in the scent, more subtly evolved than our books, that lingers in the heart of a lily; or in the wink of an eye, heavy with patience, serenity and mutual forgiveness, that sometimes, through an involuntary understanding, one can exchange with a cat.

A second sense of self-transcendence is what I will call *Education*. Somewhat eccentrically generalized perhaps, I will say that the point or end of education in the deep sense intended here is "mastery." By this I mean the self's intention to master its relations with all contexts through both internal and external discipline – to know itself by means of its ability to master the arts of dwelling in the worlds around it. I would assimilate to this category your beautifully worked out examples of the wilderness, fly-cast fishing, and the culture of the table, adding to these space travel insofar as it is devoted not to new knowledge but to learning how to navigate the universe. Be it through the roar of rockets, the silent whisper of fly lure on water, or the measured dance of a stately meal, we pursue focal things with an educational motive if our concern is to negotiate the permeable boundaries of self and world; to master the arts of making ourselves "at home" both in natural and human settings.

Finally, there is the self-transcendence of *Frontier Crossings* in quest of authentically new knowledge. Do we not cross frontiers in order to move beyond the boredoms of the known, in tiny increments to approach the pattern of God's own mind? One peculiarity of this way of looking at things bears notice. Once a frontier has been crossed, either in the sense of research or the rigors of exploration, the knowledge that it can be means that subsequent crossings come under the heading of education; the mastery of making oneself at home across the new frontier. Self-transcendence through educative mastery rescues from ennui the second crossing of any frontier. Why else would anyone seek to "discover" twice the source of the Nile, or run yet again the four minute mile? The word "adventure" often obscures the differences between these two projects of self-transcendence. Of course, both projects can be captured within a single adventure (e.g., the career of David Livingston). But the crossing of a frontier changes the world forever for those who come after the event. In the economics of self-transcendence, it may thus be said, each frontier crossed adds to the fund of available educative adventures, but this fund of adventure will increasingly lack the possibility of authentic discovery. Deep awareness of this phenomenon is almost unique to the twentieth century and we have yet to ascertain its evolutionary effects upon world culture. The possible bearing of this observation upon our population's fascination with drugs would be interesting to explore.

Now, it seems to me that contemplation, education, and frontier crossings reflect diversities of orientation toward experience that lie beyond good and evil and are, in some of their instantiations, incommensurable. The detached mystic, the imperious cosmopolite, and the Faustian student of reality are different social types at heart. They differ not just in their subjectivity; the world presents itself differently to them. Deferring to each others' view of things comes hard, too hard I think to be reconcilable by enthusiasm, sympathy, and tolerance. What passes between such folk can as easily be impatience, contempt, and fear. To see this, I think we need to move beyond the examples of focal things you give us into some that illustrate less benign effects of enthusiasm. Examples, in other words, of focal things organized around darker versions of ends or means.

Evil:

War, power, empire. Are they not focal things and practices? Indeed, does not modern technology threaten these with displacement as it has other focal things? Certainly the character of modern weapons, as Jonathan Schell dramatically reminds us in *Fate of the Earth*, has rendered war and its supposed virtues pathetically obsolete. Technologies of surveillance, communication, and weaponry similarly threaten to rob the exercise and pursuit of power of its heroic associations, placing it within reach of those who need but purchase its devices. In such a world, of what point is the quest for empire? As the novels of John LeCarré suggest, and the career of Kim Philby incarnates, even these dark nobilities are fated to be robbed of their romance. What strikes me, Albert, is that your lament for focal things and practices leaves out examples which threaten your equation of enthusiasm, sympathy, and tolerance. "Enthusiasm gives deictic discourse the force of testimony. Sympathy requires that one testify not simply by setting out in some way what matters but by reaching out to the peculiar condition in which one finds the listener, by inviting the listener to search his or her experiences and aspirations; and so one ensures that the listener is as fully engaged as possible by the concern to be conveyed" (p. 178). No one knows that better than the founders of the cults which, in their enthusiasm, ravage our people with intolerance and subvert what remains of their cosmopolitan sympathies.

Beyond that, enthusiasm seems increasingly to take the form of Manichean dualisms and terrorist messianism. In such a world, sin itself becomes degraded into naked evil; from T. E. Lawrence to Abu Nidal, as it were. If you would redeem our anomic landscape of rootless commodities and disembodied thrills by way of a restorationist program of focal things and practices, why do you think you can avoid the pluralism of humankind's divided soul: contemplation at the service of ennui as well as beauty, mastery in the name of empire as well as peaceful abode, and frontiers breached in frenzied defiance of all limits as well as in pursuit of reason?

Conclusions:

So, my friend, I cannot go all the way with you in your assault upon the citadel of technology. You have not, I fear, shown us the way to a polity of goodness so much as you have shown us the heart of a good man. Where do I stand in relation to our common project of holding the line against technicism? More with Rawls than you do, it seems; with his view that distributive justice, not goodness, must remain the first virtue of social institutions. I hold to this position despite your arguments (pp. 96–97) about technology's distortion and assimilation of Rawls's meaning of "opportunities." I have not found in your replies to Rawls, any more than in my own writing, persuasive cause to reject my model of the "libertarian-technicist society," to which you refer, as a reasonable account of a just society. To be sure, it is not one you or I prefer. Yet it takes account of the deeper pluralism of focal things and practices which, as I have said, I do not think you sufficiently acknowledge. True, like the wilderness, the precariousness of the multiple worlds in such a society is based on the rule of technology which has deprived most of them of their primordial innocence of authenticity. With modern technology, the possibilities of focal things and practices based on the quest for new frontiers to cross recede from the grasp of ever more people, leaving largely those practices rooted in contemplation and education. About that ratio probably little can be done. Further, some projects based on education (such as war, empire, and perhaps much of true politics) must have their scope of focal practices restricted in the name of public safety. There is some danger that contemplation will become the dominant mode of self-transcendence for masses of people in future versions of the technological civilization. These thoughts suggest the need for more systematic analysis and classification of focal things and practices under modern conditions than any of our present treatises on technology and culture have taken account of. Perhaps this is a frontier for future collaboration among us, toward which your book splendidly points the way.

In any event, you argue that "even when technology is explicitly envisaged as a social problem and in its connection with democracy, technology has a tendency to deflect a searching examination" (p. 97). The pluralistic justice of a libertarian-technicist society is not deflected by technology in our time, I think, so much as by capitalism's need for labor power and the consequent persistence of late industrial capitalism's version of the Protestant work ethic. Two decades ago, a more imaginative era by far, Robert Theobald (in *Free Men and Free Markets*) suggested liberating the true energies of the market principle by means of a constitutionally guaranteed minimum subsistence income. This, he reasoned, would sever the connection between survival and the vicissitudes of the capitalist labor market. Once guaranteed survival, people would organize themselves into producers and consumers according to principles closer to what you mean by focal things and practices (including politics) than is now possible. Freed from the drag of union interests, industry would cybernate thereby making possible the conquest of material scarcities of many sorts. Theobald never adequately asked why such schemes do not bear experimental fruit; for that he would have had to turn to questions of institutional arrangements, power, and privilege. Until something like such reforms are attempted, we will not know which constraints on experience are due to technology and which to the institutional conditions of its deployment.

Your great contribution in this book is to remind us that life is ultimately about the qualities of experience; that it is to the status of those qualities that we must look for criteria for social criticism; and that technology as such (or technological language which was the topic of my book) can never safely be regarded as merely neutral. As I look past this era of impoverished imagination, I still find in the call for a greater justice of primary goods distribution, with its implicit demand for more humane dispositions of technology, the main short-term hope for improving the

common lot. In the longer term, there may indeed follow an evolutionary period of pluralistic experimentation with focal things and practices
of the sort I have called the libertarian-technicist society. Perhaps only
through such a restricted experiment with primal liberty will the human
race discern a unity within plurality. It is to this long-range critique of
the present that I think your book has contributed much. For whether it
be technology, capitalism, or just plain human frailty which is the root
of our malaise, your meditative and musical philosophical essay recalls
our attention to "what is needed if we are to make the world truly and
finally ours again . . . the recovery of a center and a standpoint from
which one can tell what matters in the world and what merely clutters it
up" (p. 225).

Syracuse University

SYMPOSIUM ON ALBERT BORGMANN
III. Reply

I am grateful for your generous and provocative remarks and even more so for your expressions of friendship.

Although all of your critical points raise important problems, I have decided not to answer them one by one. Rather I have tried to understand their common force and to rethink the issues on which they converge. There are, not unexpectedly, two such issues. The first amounts to the question whether the device paradigm of technology discloses the central difficulty in the character of contemporary life. The second asks whether the notion of focal things and practices leads us to a clear and secure reform proposal. In answering these questions, I will use the concept of public moral discourse as a connecting thread. In the first part of my remarks, I will discuss a certain kind of Marxism first to set off public moral discourse from structural social analysis and then to determine the appropriate topic of public moral discourse. In the second part I will try to show that focal concerns constitute the appropriate origin and end of public moral discourse and how the alternatives and counterinstances suggested by Professor Carpenter and Professor Stanley fit into this view of moral discourse. In an extended parenthesis I will note and attempt to illuminate misleading and inherently difficult aspects in my account of focal concerns. And I will conclude with some brief reflections on technology and evil.

To begin then with Marxism, I grant that my treatment of it is unsatisfactory. Like Professor Stanley, I was unwilling or at any rate unable to master this vast body of scholarship. I apologized for this in the book and tried to indicate why I had discussed Marxism anyway. "Clearly, I have not done justice to the range and subtlety of much of Marxist work," I said. "My concern has been roughly to attend to the Marxist claim regarding the social center of gravity and responsibility." My claim was "that Marxists misplace that center and that an analysis of technology allows us to locate it more appropriately."[1] I accept Manfred Stanley's reply that the claim is too sweeping. I do think it holds for the writings of Baran and Sweezy, Braverman, Domhoff, Ewen, Habermas, Leiss, and Marcuse that I considered. And it also holds of the crude

29

Paul T. Durbin (ed.), Technology and Contemporary Life, 29–43.
© 1988 *by D. Reidel Publishing Company.*

Marxism that appears in many versions and contexts. For convenience, let it be understood that I mean the writers just named when I speak of Marxists in what follows.

But whatever the doxography, the crucial and substantive question regards the nature and significance of "the social center of gravity and responsibility." The search for such a center springs from a dissatisfaction with technology and a desire to reform it. To change the course of something, you must know its center of gravity; to change social conditions, you have to know who or what is responsible for them.

Let me present my view by using contrasting positions, drawn from the Marxists I mentioned, but also from cruder forms of Marxism and from liberals. These contrasting positions, I believe, are valuable and perhaps indispensable to articulate the position that concerns me. But even if this is so, Professor Stanley's question remains whether a consideration of the higher and wider Marxist literature would not have led me to a better or even different placement of my view.

Why then do these Marxists believe that a change in this country's social order is needed? It is because of the inequality of the present order. What is it that needs to be equalized? It is basically economic power. What is needed is equality in the shaping and direction of the means of production and in the enjoyment of the fruits of cooperation. I share this call for equality.

Then the next question is: Whom are we speaking for when we call for equality? Marxists and many liberals believe that all but the rich and the powerful desire equality of this sort, and that in calling for equality we are speaking for most people. This, I believe, is empirically false. The poor and the powerless in this country and around the globe do desire equality. But the broad middle class in this country does not. The fact that it does not has a crucial bearing on the call for change. A call for reform is futile when those who alone are in a position to change things remain unmoved.

There is good evidence that the middle class of this country does not desire equality. The evidence lies not only in the fact that the middle class has failed to support strong egalitarian economic measures since the beginning of the republic. It also appears from polling data. To be sure, the lack of commitment to equality is a complex matter. A forceful illustration comes from Reagan's reelection. Two months before the election certain voters disapproved of Reagan's handling of the economy and of foreign policy, had the same views as Democrats on most issues, felt the Republicans were more conservative than themselves on most issues, felt that Reagan sided with special interests and Mondale with the average citizen, felt that Mondale's programs would be fairer to all people and that Mondale would reduce the threat of nuclear war more than Reagan would, and finally felt that blacks had been hurt, not helped, by Reagan. These voters, if any, seemed to desire greater equality and should have been expected to lean toward Mondale. But by a majority of about seventy-five percent they preferred Reagan anyway. Why? They had one thing in common. They all felt that they would do better financially under Reagan.[2]

This particular poll illustrates two points that are evident in many other ways. The first is that the middle class in this country is divided on the issue of equality or, more broadly, of social justice. There is a sense of fairness and of compassion and an inclination toward greater equality. Hence proponents of equality believe that they represent the aspirations of a majority of the people. And this very often gives them a sense of mission and power. But such a sense, however rewarding momentarily, is ultimately futile and impotent because it ignores the ultimate and overriding commitment of the middle class to economic advancement.

This debilitating delusion of the Marxists is more a matter of rhetoric than substantive positions. They often write with a sense of certainty and vigor that presupposes the approval and support of all but the very rich and powerful. More important, most academics in my experience reach for that same rhetoric of powerful egalitarian indignation when they speak to controversial issues of public policy. But all these endeavors remain ineffective in the end because they fail to address people in their deepest conviction. What concerns me is not so much the empirical or philosophical error that manifests itself. It is rather that so much well-intentioned and needed commitment to change is misdirected and wasted. (Let me add parenthetically that my claim about the Marxist rhetoric is not easy to substantiate and requires an extended stylistic analysis that I cannot provide here.)

Of course, it has not escaped the attention of the Marxists that the people in this country have failed to act on their sense of equality. The Marxists see them as lying in bondage to inequality and therefore in need of liberation. With this too I agree. But I cannot accept the Marxist answer as to what keeps them in bondage and what is likely to release them.

The great difficulty here is to obtain a helpful view of the relation between people and institutions. On the one hand we must acknowledge the intimate ties between persons and their institutions. The view that sees human beings in their most essential and moral respect as distinct above and beyond their language, culture, and environment must be rejected. If on the other hand, persons have no independent standing at all *vis-à-vis* their present institutions, reform will be impossible. For better or worse, neither the Marxists nor any of us have the power to liberate people by changing the institutions. We can only hope to appeal to people and to change the institutions through them.

Let me repeat that it is naive to think of people as unencumbered by institutions and as approachable in a transcendent sort of freedom. But this naiveté is no more debilitating than the sophistication that exhausts itself in structural analyses of society and loses persons as so many strands in the total social fabric. The debility is not one of inconsistency or triviality. There is nothing self-contradictory in describing the social reality as an intricate structure in which individuals are but partial and dependent features; nor need it be the case that such descriptions are trivial and unilluminating. But clearly such a view fails to disclose and distinguish human beings as privileged bearers and agents of significance and therefore of reform.

Now if we do recognize human beings as the eminent points of the social order, as uniquely capable of grasping significance and of making it prevail, and if we want to reach them in their unique potency, then we must employ a particular kind of discourse, one that is, broadly speaking, moral. If we want to reform, we must moralize. And the Marcuse of *One-Dimensional Man* certainly speaks in morally charged ways. So do the other Marxists I know. But they do so ambiguously. Many of their arguments can be read either as a structural analysis or as a moral indictment. This ambiguity is a familiar Marxian bequest.

Since I am committed to a reform of the present social order, I consider the moral reading of Marxists as decidedly superior. Let me trace some of its crucial features and also stress that such a reading brings out only half, if the better half, of their arguments. The resulting picture is familiar. In it the bondage of inequality is seen as being inflicted on the workers by the capitalists. The capitalists are the perpetrators, the workers are the victims. The workers can overcome their tormentors if they see them for what they are. It is the task of the Marxists to open the workers' eyes.

One can rightly take offense at the crudeness of this sketch. Matters are much more complex and subtle, and so are the Marxists' arguments. All this I concede. But I want to stress again that if the social analysis is complicated to the exclusion of the capitalists and workers as moral agents, as bearers of responsibility and agents of reform, then I would find such an analysis of much more limited interest.

If we grant the moral reading of the Marxist work and also grant it the circumspection and subtlety that it in fact possesses, then the Marxist account amounts to a message of liberation and of prosperity. Such good news should spread fast and far no matter how initially limited its exposure. But obviously it has not. Barred from the retreat into an amoral structural account, the Marxists face two possible explanations. The first is that the overwhelming power of the capitalists prevents the workers either from acting on their vision of liberty and prosperity or, worse, from preserving and maintaining the moral integrity needed to grasp that vision in the first place.

This is the explanation that is suggested by the Marxists that I am familiar with. It amounts to what I have called "the massive exculpation of the people." Professor Stanley has shown me why this way of putting the issue is a partial and misleading critique even of the few Marxists' writings that I have considered. I took a moral reading of these texts without first distinguishing it from the equally or perhaps more plausible structural or so-called scientific reading and without then making a case for the superiority of the moral reading. "Exculpation" has an unhappy ambiguity in the book. It can mean either that the people are declared free of responsibility because the issue is not a moral one to begin with but rather a structural one or that they are declared free of responsibility because in their case ought fails to imply can.

Here too I frequently find the Marxist position the last resort of academic liberals. I do not just mean my friends in philosophy, sociology, and political science, but those in literature, history, jurisprudence, and in the sciences as well. These are primarily my people whose culture, dissatisfactions, and aspirations I share most immediately. Their education, tradition, and temperament along with the relative simplicity of their life, sometimes voluntary, sometimes imposed by a board of regents, make them eminently able to grasp and to advance the reform of technology. But too often I see their political insights take a finally futile turn when they bury their hopes in the exculpation of the people.

Still, I agree with the Marxists and the liberal academics that people are in bondage to inequality and should free themselves of it. And since I have claimed that they could but do not, I should explain why they do not and why they should. What is also needed is an explanation of what is meant by "moral." My critique of the Marxists seems to employ that naive separation of morality from social conditions that I had rejected earlier.

Let me begin with morality. I use it in a broad sense to designate the concern with and for the common order. We are concerned *with* the common order when we analyze and describe it. But in all such analysis and description we also have a concern *for* the common order. We criticize it, defend it, attempt to change and reform it. We take responsibility *for* it. The moral order, as Professor Stanley has shown in his book *The Technological Conscience*, is always a "forensic construction."[3] Something crucially valuable is always at issue in the common order that calls on us to defend it, to clarify it, or to help its being born. Particularly in the classical social theory of the modern era there is, as Professor Stanley says so eloquently, a "persistent hope for progressive redemption from the present."[4]

I have claimed that the present from which we need redemption has since the Enlightenment been shaped by a promise of liberty and prosperity and taken on a peculiar and pervasive pattern that can be explicated through the device paradigm. Technology is the kind of life in which this pattern is dominant. How are people related to this ruling pattern? They do not dominate it as though technology were a mere or neutral instrument. Nor are they dominated by it as though it were a substantive and overpowering force. But is there a third possibility? I have suggested that there is and have called it implication in technology. We can distinguish an ontological and an epistemological aspect of implication. Ontologically implication means that we were all born into a world that always and already exhibits technological character. Prospering or simply surviving in this world, we have affirmed it to a large degree. We are inevitably implicated in it. Epistemologically implication means that we have a profound and implicit understanding of technology. We understand its basic structure, and in particular we understand its promise and rewards. I would have been much less confident of this suggestion had I not been able to rely on a section of Professor Stanley's book entitled 'Sociology and the Sources of Legitimacy' in which he surveys and explicates theories that countenance a twofold commerce that we have with reality, one at a profound level where we have agreed on a basic mode of social practice, and another at the surface where we go about our daily tasks.[5]

In the most common explicit view, technology is seen as a mere instrument that we employ at the surface of daily life. In urging that technology constitutes the deep structure of modern life, I may have spoken of it in ways that suggest a substantive or determinist view of technology. To be sure, technology does determine the surface; but it does not determine and overrule us at the deep level. I use the terms "consumer" and "taxpayer" to designate people's surface roles. In these roles we have to act out a script, and we can be excused for speaking our lines. But the excuse is always conditional because as citizens we are responsible for the script or the deep structure, and from this responsibility we cannot be excused. If the deep structure that we have agreed to yields the need for two jobs in a family or the loss of work through unregulated plant closings, we have no cause for complaint.

Unlike the Marxists I am familiar with, I believe that most of us have freely accepted the present deep structure of society. What does it mean to be free at that level? It simply and only means that one is capable of an alternative fundamental approach to reality. The word implication suggests something like this. To say that you are implicated in a situation means that you are caught in it, but not altogether ignorantly and helplessly. Our implication in technology manifests itself in our ambivalence toward technology. And there is at least some empirical evidence to support the claim of implication and ambivalence although the usual polling and questionnaire methods are not really sensitive to social deep structures.[6]

How is the implication in technology related to the bondage of inequality? I think the bondage is freely entered into. It is the kind of self-incurred tutelage of which Kant speaks in his essay on the Enlightenment.[7] But why would people freely subject themselves to an order of inequality? Inequality in the Western industrial democracies has been fused with technology in a way that serves to strengthen technology. And it is technology that people have bound themselves to, not inequality as such. And as regards technology, people are not mistaken about it. They do not have a false consciousness of it. They have a correct if implicit understanding of its promise and its rewards.

This claim needs qualification, to be sure. I have already said that people's explicit and superficial understanding of technology is commonly mistaken; and if they were to confine their awareness of technology to the technological surface, they would have a false consciousness. But I do think that people have a complex appreciation of technology, one that is superficial and profound at once. This complexity comes to their attention privately in occasions of decision and also in public forums of decision.

Moral discourse is most importantly about the deep structures of society, more particularly when it questions the presently prevailing structures. This is also where we most decisively appeal to human freedom. Such an appeal speaks to people in their prevailing conditions, but it gains its force in speaking on behalf of an alternative common order. People's ambivalence about technology is one indication that they are free to appreciate the force of an alternative. It depends both on people's appreciative capacity and on the force of the alternative whether they will free themselves of their present bondage. The Marxists' and liberals' appeal to equality is not forceful enough to make people revoke their agreement with the present order. What is crucial, I believe, is the substance of the kind of life that we are to share equally.

Let me quote two passages in which Marcuse in *One-Dimensional Man* sketches the substance of the good life.

The technological processes of mechanization and standardization might release individual energy into a yet uncharted realm of freedom beyond necessity. The very structure of human existence would be altered; the individual would be liberated from the work world's imposing upon him alien needs and alien possibilities. The individual would be free to exert autonomy over a life that would be his own. If the productive apparatus could be organized and directed toward the satisfaction of the vital needs, its control might well be centralized; such control would not prevent individual autonomy, but render it possible.

And somewhat later in the same book:

Advanced industrial society is approaching the stage where continued progress would demand the radical subversion of the prevailing direction and organization of progress. This stage would be reached when material production (including the necessary services) becomes automated to the extent that all vital needs can be satisfied while necessary labor time is reduced to marginal time. From this point on, technical progress would transcend the realm of necessity, where it served as the instrument of domination and exploitation which thereby limited its rationality; technology would become subject to the free play of faculties in the struggle for the pacification of nature and of society.[8]

None of the Marxists is much more specific than this nor is Marx himself in the *1844 Manuscripts* whose rhetoric is clearly audible in Marcuse's remarks. Marcuse nowhere in his book provides a significantly more detailed positive account of the good life though he gives it an extensive negative elaboration by using it as a critical tool. In its positive features, it seems vague and utopian to me and for that very reason technologically specifiable. This kind of life is not essentially distinct from the one that is supported by the technological deep structure. And hence, apart from its egalitarian cast, Marcuse's vision of the good life has no deep critical force.

One might object that equality and the substance of the good life are not separable from one another. This, I believe, is correct in one sense but not in another. It is correct that a particular kind of life, say technology, gives rise to its own kind of equality and inequality. Technological equality is different from monastic equality, social inequality differs from communal inequality. Hence one could criticize technology profoundly in the name of, say, communitarian equality. But the Marxist concepts of equality that I know are compatible if not of a piece with technology. And therefore one could meet Marcuse's demands by making technology egalitarian. But such a program, as I have urged, has no appeal for most people in this country.

What then would appeal to people? This is the question of the second part of my reply. My suggestion in the book has been that focal things and practices would. Both Professor Carpenter and Professor Stanley have cast doubt on the social and political reforms that are inspired by focal things and practices. Professor Carpenter has suggested that monetization rather than technology has been the negatively transformative force in the modern period and that alongside it the civic culture has been a moderating and benign counterforce; and he has implied that we should look to it rather than focal concerns as the agent of reform. Professor Stanley has urged that the resort to focal practices may not only be unnecessary but tragically misguided in that it may open the door to fanaticism and terror. The liberal technological society, he suggests, may be well advised to keep the lid on Pandora's box.

Both positions have had strong support in recent debates. Let me turn to Professor Stanley's position first. It has been defended in reply to Michael Sandel's attack on liberal democratic justice. Brian Barry made the point most forcefully:

Sandel makes the transcendence of justice by group identity sound very high-minded. But it gives the green light to every string-pulling parent and crony-hiring academic. And at the end of that road stand Torquemada, Stalin, Hitler, and Begin. Sandel's argument should be turned on its head: it is exactly when "devotion to city or nation, to party or cause" run deepest that the constraints of justice on the pursuit of those allegiances are most needed.[9]

This line of defense, though heavily occupied by Barry, Rawls, Dworkin, and others, is much weaker than is commonly thought, and it is not Professor Stanley's line. The supposed strength of that first line of defense is its moral neutrality, the freedom and openness it seems to guarantee. But Professor Stanley himself has done much to reveal how morally charged and consequential the present order is. It is clearly not the best, but he thinks it may be best at keeping the worst at bay.

To this Sandel and others have replied that the present version of liberal democracy not only fails to be the best but also fails to be self-sustaining and must in hidden ways draw on the very best while simultaneously destroying what it depends on.[10] One of those sustaining sources, we might say with Professor Carpenter, is the civic culture. In *Habits of the Heart*, Robert Bellah and his collaborators have called this the republican tradition and strongly urged that we acknowledge and nurture it.[11] The wide and warm reception of the book shows that Bellah has struck a responsive chord.

But it seems to me that neither the liberal defense nor the republican reform hold any promise without invigoration by deeper concerns. To begin with liberalism, it is uneasily poised between resignation and promise. The liberals rightly accept and praise our society when holding it against totalitarian regimes. But they accept it only to the extent that it represents their ideals, and to a large extent it fails to do that. No liberal could be a defender of the *status quo* across the board. The liberal defense of the present order always carries a promise of greater justice. But for a generation now the progress of liberal justice has been marginal in this country. And internationally this country has if anything retreated from a position of justice. Thus the liberals are torn between the unacceptable resignation to the *status quo* and the discredited promise of liberal justice to come.

The appeal of the liberal ideals to the extent that it goes beyond technology has been exhausted, and it is this liberal exhaustion that has shifted the public and scholarly interest to the conservative side. The conservatives, I believe, are willing to reach deeper into our moral sensibilities and aspirations. But so far, I believe, their endeavors have been inconclusive and inconsequential. Take, for example, their recommendation of the republican virtues.[12] Such recommendations will not be accepted on someone's word. There must be an articulate and eloquent force at the origin of such a word, and any such word must point to an end that speaks for itself. It was my proposal that focal things and practices are the proper origin and end of public moral discourse.

Let me try to make that suggestion plausible by elaborating it against the particulars of Professor Carpenter's and Professor Stanley's objections. I begin with Professor Stanley's suggestion that the notion of focal concerns is both too narrow and too wide; too narrow in excluding inclusive ends which are evidently valid and admirable counterpossibilities to technological shallowness; too wide because it fails to exclude pernicious practices. In replying to the concern about narrowness I would consider the inclusive ends of contemplation, education, and frontier crossing as important extensions of focal concerns, not as alternatives. These inclusive ends require things and practices for their metatechnological identity. If for instance contemplation fails to take the form of a contemplative practice and is left to the vagaries of remarkable things drifting into my life, then most of my life will be spent according to the prevailing pattern of technology, and my contemplative capacities will at length atrophy. Conversely if the mastery taught by education fails to be the mastery of a thing that is deep and rewarding in its own right, if, for example, it is mastery of a Nautilus machine or of a bowling alley, then mastery will become vacuous and seek its rewards outside of itself, in narcissism or in the purses of sponsors.

What about the problem of evil, however? Is it not clear that the concept of focal things and practices allows for destructive and unacceptable instances? The answer is that in the sense of this question I have never proposed or developed a notion of focal things and practices, i.e., a concept that can be abstractly defined as, say, a set of necessary and sufficient conditions. I have talked about particular focal things and practices, about the wilderness, running, and the culture of the table. These, as Professor Stanley said, are self-justifying or self-warranting. They are not warranted in meeting transcendent criteria. I did say that as a matter of fact there is a kinship among these different things and practices that can be explicated as a set of generic features. Such a set is the recollection or anticipation of a focal concern. But finally the focal thing must warrant itself to us, and I am sure that terrorism would fail to do so.

In spite of the explicit provisos in the book, there is something incomplete and misleading in it. I did after all give examples of focal concerns. If they can all coexist in a person's life, then none of them can have the uniquely focusing and centering force that has been ascribed to them severally. Rather they require a higher harmonizing principle such as a Rawlsian inclusive end. If they cannot coexist in one life, they obviously must have been presented as different versions of one and the same thing, and that thing can only be an abstract defining or constitutive concept. It is natural for the reader to resolve the problem in the latter way.

The book argues for particularity and uniqueness. But it is inconsistent in avoiding final uniqueness, in being indirect about the author's focal practice, the one that finally matters. I have meanwhile dealt more directly with this concern in a number of essays in which I have tried to explain my precarious Catholicism.[13] It was not, however, an oversight or just excessive reserve that led to this confusing indirection. The difficulty that one faces here is deeply ingrained in our profession. It is an ambivalence that mirrors the wider cultural ambivalence noted earlier.

Reliance on particularity, i.e., on unique things and events is on the one hand inevitable in moral discourse and on the other hand suspect and commonly disavowed. The reliance is evident in ethics from the widely observed fact that all moral principles that have ever been put forward, from Platonic ideas by way of Mill's utilitarianism to Rawls's principles of justice, are subject to vitiating instances. The cheap response to troublesome cases is to tell the sceptics that they have misunderstood the principle, that the counterinstances stem from a category mistake. The more obliging reaction is to elaborate or tighten up the principle so that it will henceforth exclude the counterexamples. We know how the story continues. The more elaborate principle remains vulnerable to other kinds of counterexamples. Moreover it begins to exclude cases that should be sanctioned. One finally realizes that the principle can be fully explained only by presenting a case that instantiates it positively and eminently. You have to tell a story. And this is to say, as Alasdair MacIntyre does, that the narrative is the discourse appropriate to ethics.

Moreover it is a narrative that you can vouch for; it should be your story. But at this point reluctance and reticence overtakes the professional philosopher. To expose one's deepest personal convictions to public scrutiny seems indecent; it is anecdotal at best and special pleading at worst. We seem to be hopelessly dispersed into our individual concerns, and it seems pretentious to say otherwise.

But I think philosophers are wearing professional blinders in this regard. There are authors who write about their life in particular, who know it to be a good life and are not deterred by our moral diaspora. They speak confidently about what concerns them focally. Their pedagogical intent is indirect but powerful, and they possess a narrative gift I do not have. So instead of speaking of my focal practice, I appealed to them as witnesses for concerns each of which I have shared to some extent. It seemed to me that in speaking of the good life, one faces two extremes. One can try to invite everyone into the conversation by demanding next to nothing of anyone; or one can demand much and speak to rather few. The first approach is that of John Rawls; the second is that of John Richard Young who will not speak to you unless you are ready to buy a flat saddle for your horse with the stirrups hung close to the middle of the seat.[14]

I wanted to attend to John Richard Young's side of the spectrum but in a way that would show him as less isolated than he appeared, that would disclose ties of kinship between his concerns and those of runners, artisans, poets, and others. Just why there are these generic bonds and how far they extend needs more inquiry. The first task, at any rate, is to bring them out, and the notion of focal things and practices was intended to do this.

The notion as presented in the book is probably too sharply drawn, especially in its emphasis on the uniquely centering power of a focal concern. People live not only in the circle of one focal thing but also in ellipses of two focal points or in an array of several focal things, somehow united in one practice. Professor Stanley's remarks on history, contemplation, education, and border crossings show the need and the openings for such investigations.

In any case, focal concerns need no proposals or experiments. They already exist vigorously and variously. What they do need is more recognition and social space. To make more room for them in society is not to open the floodgates of passion with unforeseeable and potentially disastrous results. The literature of practice shows us that focal concerns constitute a steady and salutary stream in American life, one that flows along the margins of public attention and is in places overgrown with misconceptions. But we can tell what it is like and what it would be like if we provided more open and generous surroundings for it.

To establish those conditions, experiments to be sure are very much needed. The direction of those experiments must be defined however by a clear view not only of what needs nurture but also of the force that has pushed focal things to the periphery and threatens to suffocate them. I agree with Professor Carpenter that monetization is a pervasive and trenchant cultural factor in the modern period. Money is the means of all means, the physically most attenuated and powerful agent of change. Taken as a topic of cultural analysis it discloses well the radical transformation of the institutions most closely associated with it. But seen as a means of exchange it can engender the impression that the new exchanges are remarkable or questionable but what is exchanged for money remains traditional or unproblematic. Thus attention to monetization may overlook some of the concealed transformations that can be exposed through the device paradigm.

In conclusion I must say a few words on technology and the problem of evil. Even if one were to accept everything I have proposed, the thorn of Professor Stanley's question about evil would remain. He has actually been gentle in reminding me of the problem. He and I are in different ways heirs to the legacy of the holocaust. The holocaust occurred in the midst of mature technology. It would be fatuous to dismiss it as a regrettable accident, unrelated to the development of technology. The awareness of the holocaust should be part of the maturity that Professor Stanley mentioned. What I have said in the book and just now seems to be untroubled by the recognition of "the possibility of dark ends," and therefore it may appear untrustworthy and naive.

I have certainly been too implicit on this point, and I have tried to face up to this problem in the essays referred to earlier. I cannot do justice to the problem here, but I want to give an indication of what I have learned about evil from the consideration of technology. I think it is impossible to provide an explanation of evil regardless of historical circumstances. Technology is institutionalized regardlessness and therefore blind to evil. Evil originates in the interstices of technology, in a certain kind of sentimentality for instance, as Professor Stanley has shown in a disquieting analysis of *1984*.[15] Technology has, to be sure, emasculated many vices and evils. But it has not overcome them.

But who could overcome evil? Here too an answer regardless of one's circumstances is impossible. The problem of evil is finally not a challenge to philosophy or theology, but to each of us when we are met by it. For my part the legacy of the holocaust imposes a special obligation, and what power I have in meeting it I draw from the precarious Catholicism I mentioned before.

University of Montana

NOTES

[1] *Technology and the Character of Contemporary Life* (Chicago: University of Chicago Press, 1984), p. 85.

[2] Thomas Ferguson and Joel Rogers, 'The Myth of America's Turn to the Right', *Atlantic Monthly*, May 1986, p. 50.

[3] *The Technological Conscience* (New York: Free Press, 1978), pp. 83–116.

[4] *The Technological Conscience*, p. 42.

[5] *The Technological Conscience*, pp. 101–106.

[6] *Technology and the Character of Contemporary Life*, pp. 105–106.

[7] Immanuel Kant, 'What Is Enlightenment?', *Foundations of the Metaphysics of Morals*, trans. Lewis White Beck (Indianapolis: Bobbs-Merrill, 1959), p. 85.

[8] *One-Dimensional Man* (Boston: Beacon, 1966), pp. 2 and 16.

[9] Review of *Liberalism and the Limits of Justice*, *Ethics* **94** (1984): 525.

[10] 'The Procedural Republic and the Unencumbered Self', *Political Theory* **12** (1984): 81–96.

[11] *Habits of the Heart* (Berkeley: University of California Press, 1985).

[12] See my 'Republican Virtue in a Technological Society', in C. Strain and S. Goldberg, eds., *Technological Change and the Transformation of America* (Carbondale: Southern Illinois University Press, 1987), pp. 159–175.

[13] Two are referred to in the book, p. 287, note 41. A third one, entitled 'Liberty, Festivity, and Poverty: Harvey Cox on Christianity and Technology', has been published in *Philosophy Today*, **30** (1986): 179–190.

[14] *Schooling for Young Riders* (Norman: University of Oklahoma Press, 1970), pp. 136–150.

[15] 'Orwellian Love: Political Sentiment in an Age of Terror', *Center for the Study of Citizenship Occasional Paper Series*, No. 1, Syracuse, N.Y.: Maxwell School of Citizenship and Public Affairs, Syracuse University (December 1984).

ALEXANDER BARZEL

THE CO-RELATIONAL COMMUNITY AND
TECHNOLOGICAL CULTURE[1]

I. THE CONCEPT OF CULTURE AND ITS SPHERES

Culture is the way man understands the whole of being, the way he interprets the universe and constructs his world. Hence, every new way of thinking – i.e., every new culture – requires forms of existence appropriate to it. A crisis, personal, social, or political, is the gap between the expectations of the dominant culture and the patterns of human action in the context of being-together. As the twentieth century draws to its close, a consideration of the premises on which Western culture has been founded for the last millennium and more can no longer be put off. It is imperative that the principles of social existence and the concepts of man's essence be examined. For a long period there has been a continuous decline in the intensity of interpersonal relationships. The process of atomization in the social sphere has been accompanied by spiritual phenomena, by weakening of the desire to *belong*, by fear of loss of identity and freedom, by dissolution of bonds of commitment. The main trend today is to exist one by the side of the other, with minimal interrelationships, minimal norms of mutuality, and minimal obligations. This state of affairs grew out of conceptual preferences and has become a ruling aspiration. However, deep within these preferences and aspirations there lurks doubt as to their appropriateness to the structure or the human condition, to the demands of the rising *technological culture*. This doubt presents itself not only in frequent manifestations of dissatisfaction and reaction, but also in the fear that the ever-growing process of atomization represents a threat to man's existence – and that not solely in terms of wars of physical annihilation. To combat the manace of social atomism – of the mass mentality according to which direct interpersonal relationships can and should be replaced by the invisible threads of formal-external organization – *community* with its all-embracing and direct bonds of relations must be sought. My purpose in the following reflections is to investigate a conceptual system that would provide a basis for the establishment of communities.

45

Paul T. Durbin (ed.), Technology and Contemporary Life, 45–62.
© *1988 by D. Reidel Publishing Company.*

Here I limit myself to the clarification of the demands the new universal culture makes on the forms of social existence. However, in order to complete the picture, I shall deal briefly with a general framework for the study of culture.

Man exists in five spheres of culture at one and the same time, and within these spheres he also functions on several levels.

1. First man exists in a *substantive* sphere of culture, that is, as a unique species of the animal world with characteristics which are *sui generis* as to type and potency; he creates symbols, conceptualizes, generalizes, constructs systems; is social, historical, wordly; he is rational because relational and so on.

2. The *universal* system of culture imparts to man a dominant mode of thought, each one appropriate to a given period but in accord with an instrumental civilization rooted in the sciences. Even in antiquity there was a high degree of uniformity in the deep structures of understanding and acting, and this is even more obvious in later periods when scientific and technical achievements were widely circulated. The dominating power of universal cultures is very great and it is difficult to conceive of any possibility of evading this dominance, at least over any wide area.[2] Furthermore, the modes of thought of earlier universal cultures are absorbed in later ones in varying forms and with varying degrees of potency.

3. The *particular* system of culture is immersed in particular traditions, ethnic, religious, regional – for example, forest-dwellers, Mediterranean, Northern European, Hellenistic. Man always exists on one or more planes of particular cultures. The cosmopolitan view, that it is possible to live "immediately" on the universal level without reference to the patterns obtaining in the levels of particular cultures is erroneous. Its inapplicability has been demonstrated in many well known attempts: "Christian universalism" (which turned out to be cultural imperialism, whether by government power or humanitarian mission makes no difference); or the dream of "class universalism," which failed in both world wars[3] and has, since the 1950s, been replaced by the dream of a universalism built on particular national rebirth movements.

4. Man always exists in several *regional* systems of culture, that is, in realms of modes of behavior which are dependent on various functional and institutional relationships: the culture of the suburbs, of an officer corps, of women, of rural communities, of the campus, of

self-service restaurants, and the like. The ways of life of the proletariat, of the nobility, of fishermen, etc., transcend ethnic and religious limits.

5. Man also always exists in *periodic*, temporary systems of culture. Such are the intellectual fashions which rise and fall in different periods: Baroque, Victorian, the "Youth Revolt" of the 1960s, the culture of the cults. Such also are the systems peculiar to specific age groups: teen culture, clubs for the elderly, and the like.

As stated, this paper is concerned only with an examination of the distinguishing patterns of the new *universal* culture coming into bloom; however, I want to look particularly at the forms of togetherness demanded by this new universal culture.

II. CONSTITUTIVE PRINCIPLES AND REQUIREMENTS OF UNIVERSAL CULTURES

Universal cultures are systems of those epistemic principles that are dominant in an age characterized by scientific knowledge and technical capacity. Every universal culture has its own requirements for the construction of forms of society. These requirements are propositions which spell out the very high degree of symmetry between the *constituting principles* of a culture and the forms of social existence (including any institutionalized human action). Every form of togetherness has an index of compatibility, an index of its applicability under different spiritual conditions; from it may be drawn conclusions about how probable it is that a social construct will function under these conditions. For instance, we used to discuss such indices of compatibility as that of Islam to contemporary instrumental civilization (something Libyan and Iranian religious and political leaders still do); or the degree of applicability of Socialism to the Third World; or the compatibility of German culture to democracy; and so on. I take care here not to go beyond probability. What follows is not to be taken as a repetition of Marx's claim that interpretations of socio-historical processes can be endowed with scientific validity. The expectation that social forces, like "laws of nature," will do the work of *homo volens* is groundless. A radical change in the realm of social existence will come only if men are willing to constitute communities in full awareness of the significance of the new culture (including its index of compatibility).

Features of old cultures – whether epistemic principles or forms of existence – do not disappear totally. Some persist because of deep penetration into thought patterns, as is the case with those elements of mythological culture which emerge in any future culture in religion and symbols. Other features persist because of their fundamental indispensability, as is the case with the family (in spite of challenges in recent times). There are features which persist in *points of view*; those features retain their normative validity and their ideological power to motivate people to constitute forms of togetherness even when a new culture requires different norms and forms. This is the case with Hobbesian mechanistic atomism which was the soil that nurtured "amorphic society," the form of togetherness of functional-temporal association. This remains the model of social construction in our own epoch. The persistence of old features during the transformation to a new *episteme* gives rise to a fallacy since the old view still appears to be valid. Recently, as we have been experiencing the transition from the former dominant universal culture (the anthropological) to the new one (the technological), the fallacy is exemplified in the continued claim of atomism to exclusive ontological validity. We still see it assumed that the human essence is focused in power, that togetherness is necessarily characterized by competition, the accumulation of wealth, war of everyone against everyone. The claim goes further, maintaining that atomistic ontology presents the only authentic motivation, the only normative principle for the constitution and evaluation of social entities (with all the political, local and global, consequences of that view). Now, when it is time for fundamental reexamination, this amounts to a *petitio principii*, a begging of the question protected by the routine of hundreds of years of atomistic-egoistic attitudes in Western civilization.

III. FROM MYTHOLOGICAL TO TECHNOLOGICAL CULTURE

The first universal culture whose features can be reconstructed from echoes in modern human activity, *mythological culture*, flourished in *ancient natural civilization*. Its *episteme* focused on the conception that hidden powers reside in everything. These powers determine what exists, set events in motion, and impart meaning to a world conceived of as a cosmic organism. All the relationships between things that exist are determined by these hidden forces; this includes man, whose being and limits are his destiny. Harmony exists in the cosmic organism and is expressed in the language of symbols. Men and all that exists in nature are included in this all-embracing harmony. Things that oppose the calling to embody cosmic harmony – tensions, conflicts, defects – are interpreted as insults to the hidden powers. Since man's existence has always been dependent on what is given in nature, he has turned increasingly toward the forces in nature that operate with apparent orderliness. Thus, the seeds that sprouted in our new universal *episteme* were already to be found in mythological culture.

The second universal culture, the *cosmological*, flowered in a *middle natural civilization*. Its *episteme* concentrates on the idea that the forces of nature cause the existence of every entity, in accordance with the lawfulness that is imbedded in their very being. Man's own world is organized in a natural fashion based on laws. Everything that exists – including relations between existents – is determined by the forces of nature. Individuals constitute natural organisms, and the multiplicity of natural organisms makes up the equilibrium of the natural order – which is the same condition that was termed harmony in mythological culture. All forms of togetherness must be compatible with this organismic model since it provides the necessary equilibrium which is the sole warrant of undisturbed functioning. The *episteme* of mythological culture had penetrated so deeply into man's intellect that, for a long period of time, tensions and parallels (and in some cases a dualistic split) were maintained between the old and the new *epistemes*. Tension reached its peak in the West during the Renaissance; this defined the trend implicit in the foundations of cosmological culture – namely, to *comprehend* the dimensions of nature within a totally independent framework of *human thought*. Thus a vista was opened up for the creation of a new universal culture.

The *episteme* of *anthropological culture*, of *advanced natural civilization* (our most immediate predecessor), focuses on the idea that lawfulness in and the meaning of the universe are imbedded in man's controlling thoughts and deeds. Human experience reveals discrete existents, and complex phenomena derived from them are mechanisms which operate in accordance with the varying strength of the individual items, striving to realize the unique quality inherent in each of them. Life in this complex of inclinations is possible only through planned organization – up to the level of functional aggregates – for the maintenance of a mechanical equilibrium which is now the substitute for harmony. (Sometimes this equilibrium is an illusion since the fundamental opposition between particular forces cannot be harmonized.) In the sphere of human relations, equilibrium requires the existence of socio-political instructions which shape and motivate the joining of individuals into functional structures of togetherness. Man the investigator and the systematizer gains awareness of power. The world of man now becomes the center of interest and the sphere of achievement. New standards of relationship to nature are progressively created. By experiment and observation, new tools of thought have been developed which no longer mediate directly between thought and nature; active intervention in processes is now implied and promoted. Thus did a third universal culture grow.

The *episteme* of *technological culture* – where the focus is the *logos* of *technics* – is rooted in the conception that human thought and action are deeply involved in the creation and functioning of natural and artificial things. This new universal culture flourishes in an *open instrumental civilization*. Now human thought can invent processes for the construction of technological organisms which are wholes whose components operate in mutual dependence and according to the purposes incorporated in them for the achieving of an objective. No doubt, in former civilizations man transcended the naturally given and created entities which nature could not offer; there is no difference in principle between ploughing with huge tractors and turning the soil with a wooden stick, or between modern pills and older methods of birth control. However, the dimensions have totally changed, as has the depth of transcendence of natural processes.[4] Technological organisms possess a power and an operational scope previously unimaginable. They do not ensure harmony or equilibrium unless their construction is also accompanied by a channeling of the thought and of the act of creating a constructed organism toward a supreme planned order. In this manner, human responsibility may determine the fate of the universe.

IV. THE *EPISTEME* OF TECHNOLOGICAL CULTURE

Among the characteristics of the technological *episteme*, I would like to focus on those that are of concern when considering men's being-together.

In the new *episteme*, we conceive of reality as a total system, as integrated, and not as singular entities aggregated one to the other. The particular item is known by its belonging to a system that operates as a whole.

The dominance of the system in our pattern of thought endows the constitutive principles of systems with a decisive role. We are aware of the presence of such principles in the continuity of a system's existence. In the absence of these principles items would cease to exist in a relational network, which constitutes the reason for constructing and maintaining the system. So our thinking turns to ideal structures rather than to concrete functioning alone. We are aware of pre-experiential *norm*alness that is derived from the constitutive principle.

Technological thinking is decidedly teleological and future-oriented. We no longer accept the datum as determining our reality; we see in it the raw material for the application of a principle in order to achieve desired objectives intended to improve reality. Technological thinking is basically synthethic, with analysis meant to serve it. We think – more than before and with a new quality – in terms of composition and integration. We reason in a way that leads *from* and *to* a whole, from complex phenomena to a network of particulars. The rationale of the composite is always present in our thinking about the system.[5] We think in terms of permanence, in terms of actions which are continually repeated. The determining pattern is not a one-time phenomenon, exhausted in an instant, but one that is intended and expected to occur in accordance with a rationale and a principle inherent in purposive action toward a goal. The whole functions properly as a *co-relational system* in which the components exist in the permanent relational bonds required for its reliable functioning over a long period of time.

Since the results of technological activity have become very deep and powerful, the degree of responsibility grows. Technological thinking prescribes *limitations, control, supervision*. Contrary to the illusion rooted in the atomistic attitude that civilization (which increases the amount of instrumentation) reduces the degree of interpersonal commitment, our ramified, complex relationships necessitate *more* binding, *more* institutional ties. The result could be increasing formal-external dependence if a new *We-dimension* is not invented.

V. FROM A REDUCTIONIST TO A CONSTRUCTIVIST APPROACH

In the former universal cultures, dominant in *closed* natural civilizations, where material reality was considered as given and determined, conceptions of human togetherness were shaped according to reductionist models. The features of social existence could be understood only within the given and determined frames.

Naturism, assuming that man is nothing but nature, represents a first-order reduction, and a complete one: *everything* known by observation or given in speculation on nature is taken to be valid for human beings. In that modern version of naturism, sociobiology, the complete reduction points to the genes where alone are the features of being-together determined.

Naturalism, in its assumption that man is *like* nature, represents a second-order reduction, an analogical one. Human relations cannot be reduced totally to the laws of nature without the mediation of culture, which again functions within the limits of a closed given. The fundamental or primordial components of human and animal existence are very similar. Hobbesian reasoning, though it seems to be reductionist in the first sense, leaves the realm beyond the "state of nature" open for rational decisions based on the mode of relationships evaluated in terms of cultural concepts. (Admittedly, these are seen as rooted in a mechanistic atomism with the will to power at its core, and with the "artificial animal," the state as Leviathan, as the moving force.) The organismic naturalism of Ferdinand Tönnies assumes, as one option, to "take the chance of organismic life," but there is also the option of "trans-natural" artificial society.[6] In both of these reductionisms, atomistic and organismic, a first-order teleology is embodied; in order to maintain harmony man must aim also at the fulfillment of the natural given in concrete history.

The problems flowing from both reductionisms therefore called for a new conception of being-together. According to Marx, the primordial natural given is reshaped in *historical* conditions subject to transformations. In this manner, the re-constructed given becomes the foundation of *praxis*. Though still a case of reductionism, this is of a *derivative* or *generative* sort. Consciousness is derived from the social system immanent in history in its dialiectical progress, and consciousness must constitute new and better conditions of being-together. In this third-order reduction began the transition from analytic and aggregative thinking on human togetherness to a synthetic and integrative form.

The *technological episteme* faces an *open* civilization. The primordial given is not considered determined; it is but the starting point for the construction of new entities through planning, by the ideal confrontation of designed social systems with the requirements of the dominant civilization to which they are intended to be related. The ontology on which the new social philosophy is based is not reductionist but *con-*

structivist. The time has come for a *second-order teleological thinking*: not one of realizing the primordially given, but of realizing designed forms for the sake of desired aims. We cannot any longer consider the phenomena of togetherness as natural data – whether organic or mechanical – and we cannot consider them as freely, contingently evolving functional associations of isolated individuals. We must, rather, conceive of them as constituted systems, as integrated wholes constructed for a purpose in line with the new culture.

VI. FORMS OF SOCIAL EXISTENCE

To locate properly this form of community among the forms of togetherness – and also to clarify some terms currently used in social philosophy and social science – I will now briefly sketch a model of classification. Its principle is the *intensity* of relationships among components of a social system, both among themselves and in relation to the system as a whole.[7]

a. *Collective forms*:

The forms in this group are characterized by the fact that in them individuals are joined in aggregations, with their separateness preserved (to a large extent but, of course, never entirely, since no human individual can be unrelated). They are functional assemblages, limited in duration and partial with respect to the extent of contact and commitment. Collectives are emergent, not constructed social entities. The types in this group are: (1) *co-existence*, existing side-by-side with a minimum of bonds – e.g., tenants in an apartment house, shoppers in a store, crowds gathered for some event; and (2) *coordination*, acting side-by-side. Here the bonds are closer, but they are goal-defined and their validity is limited to the common task. A group of individuals coordinate a number of activities for a given time and purpose in expectations of benefits for the participants and in order to ward off difficulties arising from lack of coordination; e.g., work crews which settle upon job arrangements (shifts, etc.), or trade unions. The lowest degree of coordination is found among individuals waiting in line at a cash register or for a bus.

b. *Cohesive forms*:

Characteristic of the types of togetherness in this group is that the individual derives from the totality a part of the substance of his existence and relationships – while in other areas he is not part of a compact system of belonging. Just so long as a cohesive form of togetherness exists and acts, the relationships are organic in nature; the whole functions as one being and the individuals are mutually conditioned. There are several kinds here. (1) *Cooperation* means being together with one another for joint action, to carry out long-range objectives. It requires continuing ties as long as there is striving toward the goal, but not beyond. Cooperators maintain a partnership which has some characteristics of integration, but at a restricted level of intensity. (2) *Communion* is adhesion one to another for an intensive, shortlived merging of experience. During the time of the merging, integration, in which the individuals are drawn together consciously and emotionally, comes into existence. Although this type of togetherness is not intended for functional tasks, sometimes communion is planned; for example, rites, involving singing, dancing, bodily contact, organized around symbols. Many political regimes understand how to cultivate such forms of togetherness effectively. The concentrated intensity leads to an error, to thinking that communion involves man's entire personality. This illusion is ingrained in the plethora of cult movements of our time, in which people seek communal wholeness; they find only societies-of-the-moment, while historical reality passes them by. (3) *Co-ideation* means existing with a bond to one another in the sphere of ideas, but not necessarily in the practical sphere. Ideological movements are of this type, including entire religious systems and the sects within them (Hasidic Jews, Sufi, etc.). Co-ideates are distinguished by their awareness of the "we" ingrained in the individuals even when they are not together. They are goal-oriented social entities, meant to be continuing; anyone convinced of the truth of the idea by which the bond to his fellow man was created will strive to preserve over the years both the validity of the idea and the togetherness which (partly) expresses it.

c. *Comprehensive forms of togetherness*:

The types in this group are systems of existence *one-with-another*. These are organisms in the sense of inclusive integrations in which the activity of the components constitutes a unifying whole in contrast to aggregates, which are functional-mechanical combinations of individuals. The individual is a corollary of the whole and cannot be understood except as belonging to it. The relationships embrace the personality in varying degrees of intensity and are not limited to the functional sphere. Such types of social existence include communes and co-relational communities.

(1) *Communes* involve being with one another on the basis of an idea that encompasses all areas of daily existence, but where the affinity in concrete activity is only of secondary importance. Communal life is regarded as an instrument useful in attaining some ideal objective such as Messianic salvation – as with the Judean Desert Sect, whose members felt their aspirations could not be realized in the social environment of the "forces of evil" (a view echoed in a great number of modern communes in the West) – or self-realization. Members of self-realization communes believe that that ideal cannot be achieved in the chaotic diffuseness of the modern social milieu; examples here include the personalistic communes established in the last two decades in the counter-culture atmosphere of the West. Monasteries and mystic communities are also instances of this type. Communes involve integration and are constructed societies – but goal-directed ones.

Here we seem to reach a dividing line between atomistic and holistic conceptions of human nature and the essence of togetherness. In all the preceding forms a person could be a component of a system while remaining conscious of his separate individuality alongside other individuals like himself. Even in the communes that exist to foster self-realization, an atomistic and subjectivistic basis is prominent. An awareness of the "We" enters into individual consciousness, but it can also disappear from consciousness if everything is going smoothly in the community. On the other hand, in ideological communities a different attitude toward human nature and togetherness is embodied. This new attitude goes beyond mechanistic and even organicist atomism (the latter, as manifest in personalistic and in some religious communities); man is now conceived as *primarily social and related*. Here we come to the most intensive and comprehensive type of togetherness.

(2) *Co-relation* means *being-with-one-another in comprehensive mutual conditioning*. In co-relative societies the individual *derives from* the whole and is a *component* of the system to the full extent of his being and activity. The "I" exists in the consciousness of the "We"; it is a

related I, an I–We. There are two examples of this type, the family and the co-relational community as seen in the Israeli Kibbutz. Here there is total integration of systemic We–I persons existing together as organisms – and in the Kibbutz to an even higher degree than in the modern family. Kibbutzim are *not* collectives of individuals existing one by the side of the other, but true co-relatives.[8]

During the period in which the atomistic-mechanistic outlook dominated man and society, there was a gradual transition from forms of togetherness based on close relationships to forms based on loose relationships. By the nineteenth and especially the twentieth century, coordination and coexistence had become dominant. Now technological culture demands a return in the opposite direction, from the minimal bonds of coexistence to the intensive bonds of co-relational systems.

The time is ripe for the construction of *technological organisms*, organisms based on the *logos* of technical culture. Only these can face the challenges of our newly-emerging civilization.

VII. CO-RELATIONAL COMMUNITIES; THE KIBBUTZ AND SIMILAR GROUPS

The constitutional principles of the co-relational community – I take the Kibbutz as the best example – seem to correspond to the requirements of the technological *episteme*, to make the Kibbutz the appropriate form of social existence in the coming age.

The Kibbutz involves constructed and goal-oriented togetherness, designed to solve substantially human and circumstantially real problems. It is a true *integration*, a system of relationships in mutual conditioning; it is not a mechanical aggregate of individuals. It is a society with temporal *continuity*, not an association limited to the periods necessary for the performance of specific tasks. It is a system which encompasses all the modes of existence. The Kibbutz is a *heterogeneous* society, yet its homogeneity grows within the structure of co-relational existence. The Kibbutz is *involved* in the concrete technological world; it is not, like almost all other communal movements of our time, an escapist and isolationist social entity. It proposes to improve and perfect human togetherness within the framework of the dominant universal culture.

The principle of *organic equality*, equality built on the natural and circumstantial qualities of each individual whatever those qualities may

be, endows the co-relational community with the capability of over-
coming those differences which are the source of injustice – differences
which give rise to atomization and alienation. The integration of individual states of being in one egalitarian system of existence prevents that
competitive polarization which creates desperate frustrations.

Absolutely common responsibility relieves the burdens of fate that are
so damaging to the singular individual. This is important now more than
ever because of our rapidly changing human condition, which lessens
our feeling of existential security. Intense interpersonal relatedness
responds to the urgent quest for belonging in our situation of increasing
desintegration, of structures breaking down around overcrowded mankind.

The Kibbutz also solves the growing *problem of longer life expec-
tancy*. Retirement is not a matter of mere material existence, but of a
comprehensive belonging within organic surroundings and with unbro-
ken continuity. Co-relation also assures a solution for the ongoing loss
of stability of the nuclear family in an "open" society. The child's sense
of belonging to the community as a whole constitutes a certainty which
tends to fortify his personality; this provides him with consciousness of
security which is unavailable to children of broken families in an
atomistic society.

The We dimension, paradoxically, creates in the co-relational community a kind of *social anarchism*, yet without alienation and loss of
freedom. Institutionalized supervision and control of the "anomic"
machine is absent from the reality of the Kibbutz. Management and
administration are mere technical instruments.

Work as a social value is more than the traditional ontological imperative of *homo faber* or the socio-political imperative of productiveness.
The Kibbutz principle means that there is no hierarchical differentiation
among people doing any sort of job; moreover, no one is ever saddled,
permanently, with a single job. The whole community bears the burden
of all the needed tasks, which are redistributed according to an egalitarian logic. Because of this rotation of jobs (at long or short intervals), no
permanent social classes can be formed. The division of labor can never
become the division of laborers; the fruits of all work are the source of
satisfaction of all the needs of the whole community, without regard for
different kinds of work or personal efficiency.

Morally controlled consumption makes a double contribution to the
effort to overcome the dangers of the technological civilization. Admit-

tedly, affluent society creates the possibility of satisfying a great many varied needs, and the welfare state guarantees a minimum level of existence. Nevertheless, neither the affluent society nor the welfare state has succeeded in easing the tensions and the frustrations of those who are unable to achieve the longed-for level of the wealthy. Technological consumption is a rat race on one hand, trying to meet standards of status, while on the other hand trying to compensate for frustrations. The co-relational community removes the very concept of status with all its necessary symbols. It eliminates consumptional wealth as a criterion of human worth and dignity. In its place has developed a more moderate ("functional") consumption governed by a comprehensive moral responsibility. Competition and compensation in contemporary consumption-oriented society subordinate production to consumption – all the while accompanied by the feeling that the natural environment dominates. This is how the ecological crisis[9] emerged. All efforts toward an environmental ethic, toward a reconsideration of human stewardship of the earth, have been in vain. In the real world of competitive and compensatory consumption, there is no chance to solve the problem of ecological catastrophe. A true "environmental philosophy" is necessarily a social and political philosophy. The co-relational community, the Kibbutz, is the model of a new social construction which can overcome the environmental damage.

Finally, the Kibbutz is a *technological community*; that is, its structure of social existence exactly meets the demands of technological culture. Its principal novelty, as contrasted with past or present experiments in communal living, is that the Kibbutz is aimed at improving the present age from within, using the tools of contemporary civilization though in a radically transformed way.[10]

Defenders of the Kibbutz consider it the proper form for the new social structure that corresponds to the features of the technological age. It is not a romantic idyll, involving a constant experiencing of face-to-face relationships; it is a realistic way of life involving different personalities in different situations. Emotional ties are not the source but the outcome of a structure of total belonging, of total common responsibility – totally foreign to our age of atomization of society. The Kibbutz involves no lack of awareness that the world can never again do without computers, robots, jet airplanes (except, perhaps, after the wholesale destruction of a World War III).

In the real world of current history, a new society needs to be established. Our new industrial forms – including the new agriculture as an industry – urgently need a social carrier that is strong enough to serve as an alternative to capitalism's competitiveness and interest in domination. In the real world of technological culture, an adequate solution for the problem of insecurity – one that can face the dangers of mass unemployment as a result of fluctuations in an unplanned and uncontrolled (perhaps uncontrollable) market system – is needed. The Kibbutz represents constructive action for the improvement of our real-life culture – but it is also a constructive reaction against the catastrophes inherited from old and now irrelevant social traditions.[11]

Technion – Israel Institute of Technology

NOTES

[1] This is an expanded version of a paper presented at the World Congress of Philosophy, Montreal, August 1983; it gives the main outlines of chapters 5 and 6 of my 'Categories of Social Existence' (forthcoming in English; Hebrew original, 1984).

[2] It is extremely doubtful whether these neo-primitivist and escapist trends have any historical significance. Khaddafi and Khomeini, in their early "detached" days, as well as the Cultural Revolution in China, seem now like curiosities of the past.

[3] Stalin's thoughts on language (1941), together with the discussion that followed in Soviet thought, proved the end of "naive internationalism" – as well as the urgent need, at that time, of "Russian consciousness."

[4] See Friedrich Dessauer, *Streit um die Technik* (Knecht, 1958; original, 1927); also, Peter Caws, 'Praxis and Technology', in G. Bugliarello and D. Doner, eds., *History and Philosophy of Technology* (University of Illinois Press, 1979), pp. 227–237.

[5] Joseph Margolis in his paper, 'Philosophy and Technology,' at the Montreal congress, presented a picture of resignation and fear in the face of the dehumanized mechanical-analytical mode of thought pressed on us by technology – and his commentators, Marx Wartofsky and Don Ihde, echoed the same thing. To the contrary, it seems to be the case that technological thought is, *par excellence*, synthetic, integrative, constructive, and comprehensive in its ideas.

[6] See Ferdinand Tönnies, *Gemeinschaft und Gesellschaft* (9th ed.; Wissenschaftliche Buchgesellschaft, 1970).

[7] The literature over the past two decades on utopianism and communal experimentation is so vast that it is impossible to list everything. Here is just a sampling of interesting books (not to mention my 'Categories of Social Existence'; see note 1, above): P. Abrams and A. McCulloch, *Communes, Sociology, and Society* (Cambridge University Press, 1972); *The Alternative Way of Life: Proceedings of the First International Conference on Communal Living* (Tel Aviv, 1982); K. Bartolke, T. Bergmann, and L. Liegle, eds., *Integrated Cooperatives in the Industrial Society* (Van Gorcum, 1980); H. Bossel, *Bürgerinitiativen entwerfen die Zukunft* (Fischer-Alternativ, 1978); A. Cherns, ed., *Quality of Working Life and the Kibbutz Experience* (Norwood, 1980); P. Cock, *Alternative Australia* (Quartet,

1979); N. Cohen, *The Pursuit of the Millennium* (Palladium, 1972; original, 1957); H. von Gizycki and H. Habicht. eds., *Oasen der Freiheit* (Fischer-Alternativ, 1978); R. Kanter, *Commitment and Community* (Harvard University Press, 1972); F. and F. Manuel, *Utopian Thought in the Western World* (Belknap/Harvard University Press, 1979); K. Melville, *Communes in the Counter-Culture* (Morrow, 1972); A. Rigby, *Alternative Realities* (Routledge & Kegan Paul, 1974); R. Ruether, *The Radical Kingdom* (Harper & Row, 1970); J. Servier, *Histoire de l'utopie* (Gallimard, 1967); and B. Zablocki, *The Joyful Community* (Pelican, 1973). Although in many ways misleading because of an overemphasis on know-*how* rather than on know-*what*, the ongoing bibliography of Kibbutz literature being compiled by Shimon Shar of the Institute of Kibbutz and Cooperation Research, University of Haifa, can also be consulted.

[8] The word "collective," in common usage, does not meet the terminological needs even of the social and political philosophers and social scientists who use it to describe a total form of society. As a matter of fact, "collective" means just the opposite: *colligere* means to put together, and no more than that. Perhaps this could be dismissed as a terminological quibble if the usage had not given rise to such flagrant misunderstandings in the history of social thought; also, the term "collective" has accumulated a rich store of negative connotations – for instance, "collectivization." The only author I know who neatly distinguishes between commune and collective is Erich Kahler, in *The Tower and the Abyss* (Viking, 1967; original, 1957).

At the present time, there are in the world about three-hundred-fifty co-relational communities comparable to the Kibbutz.

The Hutterite Societies of Brethren – nearly three hundred communities in Canada and the United States, and one in England – have existed since the early seventeenth century. The principal (and almost the only) difference, in terms of the structure of social relatedness, lies in the ideological foundations of the two. For the Hutterite brethren, the community is an authentic religious expression of human existence; a faith in the divine presence – including a belief in ongoing revelation – is an absolute precondition of belonging to the community. On the other hand, the ideology of the Kibbutz rests on the establishment of a just society, which is the only warrant needed for a world that is good for human beings. This ideology is rooted in the Jewish worldview, which stresses man's responsibility for improving reality as it is. Judaism is basically a "sacral anthropology," not a theology. See my book, 'The Structure of Judaism' (forthcoming in English; Hebrew, 1978), and my article, 'Judaism and the Kibbutz', *Forum* 37 (Spring 1980). There are another fifty or so communities in the Western world that are really closer to the commune form discussed earlier; they lack some of the totality feature of interpersonal relationships in co-relational communities.

[9] Most of the literature in the 1970s and 1980s on the ecological crisis comes to the conclusion that individualistic and atomistic views must give way to communal ones. Individuals are not ready to accept limits on consumption. For example: "Finding ways to live in a more intensively communal way – *sharing* through communities which can be organized along various lines and to different degrees" is what is demanded, according to John Wilkinson, ed., *Earthkeeping: Christian Stewardship of Natural Resources* (Eerdmans, 1980), p. 306; cf. also 147, 170; also, R. Heilbroner, *An Inquiry into the Human Prospect* (Norton, 1974), pp. 140ff; and several essays in I. Barbour, ed., *Western Man and Environmental Ethics* (Addison-Wesley, 1973).

[10] At the end of the eighteenth and the beginning of the nineteenth century – at the

beginning of the Industrial Revolution with its increasing destruction of traditional rural communities and of the relatively harmonious relationships, in the towns, between manufacturers and craftsmen – a need for the construction of a new sense of community was felt. It is unfair to call the ideas of Fourier, Proudhon, Cabet, and especially Owen "utopian," since that term belongs properly to literary speculations about idyllic societies never intended to be set up in practice. At the very time of Thomas More, Campanella, and other (properly) utopian writers, others were establishing real communal forms of living. The ideas of these groups – notably the Hutterites, since their communities still exist – deserve rather to be called "topias," places of communal living that were actually established. Similarly, the social thinkers of the Industrial Revolution planned real embodiments of their ideas, and Engels's critique – contrasting "utopian" with "scientific" socialism – was unjust, a begging of the question. See Friedrich Engels, *Socialism: Utopian and Scientific* (1880; originally part of *Anti-Dühring*). If we employ the logic of the Marx-Engels conception of history correctly, no conclusions can be drawn from mid-nineteenth-century civilization (and its political economy) about the totally different conditions of production and consumption of the late twentieth century. (Cf. Louis Althusser's structuralist interpretation of Marxism.) One must assume that Engels himself would dialectically re-evaluate his claims about utopian and scientific socialism. Engels's objection treats all those thinkers as if they were ignoring the urgent need for reconstruction of the entire economy and framework of society in order to solve the problems of individuals and communities uprooted by industry in the new capitalist order; that is, he failed to recognize that these so-called utopias were meant to be a proper way to meet the challenges of capitalism. They were not a substitute for scientific – i.e., political – socialism. Looking back from our standpoint at the end of the twentieth century, one can at least question whether those supposed "laws of science" have led to the required socialist revolution – or prevented the stablizing of capitalism, or its reintroduction in some supposedly "socialist" countries. It is quite likely that socialist ideas and the corresponding political movement would gain more from investing means and efforts now given over to the large-scale proletarian revolution (assuming unchanged individualist attitudes and interests) instead in a communal reconstruction of society. Many confused notions based on early-nineteenth-century communities, mostly in America (which brings in further ideological problems – see Jackson Wilson, *The Quest of Community: Social Philosophy in the United States, 1860–1920*, Oxford University Press, 1968), could be replaced by constructive realities if serious effort were devoted to organizing a powerful political system capable of being the carrier of this kind of communal revolution. (The rich literature of the New Left, *circa* 1960–1972, echoes this search for a communal restructuring of society, but most of the actual efforts to establish communities followed the direction of the "counter-culture.")

Faced with the challenges of a technological culture and in the midst of a radically new industrial era – often called "post-industrial" – a new communal movement has emerged, this time from the opposite direction. Human activity is no longer to solve the problems of our civilization; rather, in place of a solution, human activity – or, rather, human Being – is to be *removed*, placed outside the new civilization and established on radically reactionary foundations. The basis of this view is that the self and self-fulfillment are endangered, that mass society destroys shared feelings, that the rationality of our technological culture is hostile to emotions, to spontaneity, to direct action, to close human ties – so to protect a healthy and humane life, we should react radically and negate the economic approach to life and should escape from the framework of this civilization and substitute an ethos of leisure and pleasure for the work ethic. Neo-primitivism, "dropping out," is preferred to any sort of participation in what is viewed as a fatal course of human destruction. Hundreds of short-lived communes were developed along these lines, to seek mainly or even exclusively solutions for individual and personal problems – with no real expectation of projecting a model that would help all of humankind in our technological culture.

[11] Many contemporary thinkers understand the difference between the majority of new communal experiences and the needed social reconstruction of technological culture. See, for instance, A. Cherns, ed., *Quality of Working Life and the Kibbutz Experience* (Norwood, 1980); I. Fetscher, 'Die Gefahr der Phantasielosigkeit und das Argument des Utopismus', in his *Überlebensbedingungen der Menschheit* (Piper, 1985); P. Goodman, *Utopian Essays and Practical Proposals* (Vintage, 1964); M. Markovic, 'Gesellschaft', in H. Bussiek, ed., *Veranderung der Gesellschaft* (Fischer, 1970); E. Schumacher, *Small Is Beautiful* (Abacus, 1974); G. Taylor, *Rethink: Radical Proposals to Save a Disintegrating World* (Penguin, 1972); and Fritz Vilmar and K. Sattler, *Wirtschaftsdemokratie und Humanisierung der Arbeit* (Europäische Verlaganstalt, 1978). Even R. Buckminster Fuller's *Utopia or Oblivion* (Pelican, 1970; original, 1969), though somewhat extravagant, is full of hints of this sort.

EDMUND F. BYRNE

THE LABOR-SAVING DEVICE: EVIDENCE OF RESPONSIBILITY?*

It is unquestionably people who decide to develop and introduce new technologies. But technology advances so rapidly now that it is no longer apparent to the uninvolved observer that human beings are actually still in control of it. Technology, it is said, is on a "runaway" course; it is becoming "autonomous" of human agency. Taken literally, this characterization of technological change leaves no theoretical room for moral if even for legal responsibility. In the absence of any more subtle qualifications, it amounts to an instantiation of the sort of hard determinism that rules out the very possibility of free will.[1]

The case for technological determinism can, to be sure, be argued with a brief full of indications that technology is "out of control." Conceded. Moreover, indications that human beings are still in control can be denied theoretical import by asserting the "it's-only-a-matter-of-time" principle. Also conceded. Yet it remains the case that *some* human beings are clearly responsible agents of technological change. By this I mean that some human beings not only consciously seek the development of certain technologies but tolerate or even intend fully anticipated consequences of the introduction of that technology. I believe that this claim can be generalized so as to apply in some degree to any technological change. But here I shall build my case only on the subset of technologies developed and introduced as "labor-saving devices."

Assumed for purposes of this discussion is that workers are in fact displaced and are thereby harmed by the introduction of labor-saving devices. To be established is the claim that in each instance some human beings as agents consciously intend, at least indirectly, the harm that does result from the introduction of so characterized technologies. In support of this claim I propose to discredit each of four counter-arguments:

(1) Only good is intended by the introduction of a labor-saving device, as witness the support of workers as well as management.
(2) If workers are opposed to doing away with work, then it is demonstrably

63

Paul T. Durbin (ed.), Technology and Contemporary Life, 63–85.
© 1988 *by D. Reidel Publishing Company.*

false that some people intend the unemployment consequences of labor-saving devices. For, management surely has a long record of encouraging and stimulating hard work.

(3) Even if some people do intend to reduce the workforce by introducing a labor-saving device, they are not acting freely but only as they are compelled to do in response to competition.

(4) Even if some human beings do intend the unemployment consequences of labor-saving, this does not prove they are acting against the interests of workers. For, the long term result of this process will be a state of affairs in which people will no longer need to work.

Even if I am able to neutralize these objections, I will not, of course, have established any universal claim with regard to responsibility for technology. But the evidence I propose to introduce with regard to labor-saving devices might at least serve as an invitation to the opposition to assume the burden of proof.

I. LABOR-SAVING AS A EUPHEMISM

(1) Only good is intended by the introduction of a labor-saving device, as witness the support of workers as well as management.

Response: Appeal to worker support is misleading, because the notion of a labor-saving device is (a) ambiguous and (b) euphemistic.

Saving labor does not have the same meaning in everybody's vocabulary. In particular, it means something quite different to the managerial class than it does to the working class. To management saving labor means reducing production costs by lowering wages and/or benefits, shrinking working hours and/or the workforce itself, by automating or whatever. Working people, on the other hand (at least those who are not politically astute), typically associate saving labor not with any reduction in employment but with a reduction in the amount of *hard, debilitating work* that they must do to earn a living and/or carry out their domestic chores.

There is, then, both agreement and disagreement between management and workers with regard to the import of "labor saving." All agree that what should be saved, i.e., reduced or eliminated, is tedious debilitating work. Disagreement, and thus misunderstanding, has to do with the relevance of such labor-saving to jobs. Workers generally favor reducing or eliminating not wages or jobs but only the demeaning aspects of work. Management, however, typically favors reducing or eliminating both the demeaning aspects of jobs and the holders of those jobs.

Evidence for this latter claim includes at least the following three observations. For one, elimination of unpleasant aspects of work is typically cited as a reason for adopting new technology.

Secondly, in the absence of new technology, management typically appeals to reasons for doing the work anyway. But to the extent that the work in question is unpleasant, the reasons cited need to overcome what is probably a natural aversion on the part of the workers. So most attempts to "sell" labor to workers depend upon some special gimmick.

Perhaps the most daring gimmick of all, utilized by Voltaire and by some German philosophers in the nineteenth century, is to declare flat out that labor is wholesome, meaningful, and fulfilling. Few defenders of labor have been quite so daring. Most have taken the softer approach of emphasizing the instrumental value of work as a means to some end. Work might be praised as the way to a better world, e.g., in the Marxist perspective. Work (associated particularly with labor) might also be recognized as painful but valuable for that very reason because of some indirect benefit that arises out of the pain. This approach has been widely utilized in the West, especially in the form of a "work ethic" that encourages people to work hard by appealing to their interest in being saved – not in this life, but in a promised life to come.

Thirdly, even jobs which are not inherently or even predominantly unpleasant are continually being eliminated when and as new technology becomes available. This in itself is a strong indication that the notion of a labor-saving device, as understood by management, is a euphemism. This, however, will become clearer from considerations below of the anti-labor bias of management.

II. MANAGERIAL SUPPORT OF HARD WORK: THE INTERIM RULE

(2) If workers are opposed to doing away with work, then it is demonstrably false that some people intend the unemployment consequences of labor-saving devices. For, management surely has a long record of encouraging and stimulating hard work.

Response: The appeal to hard work is only an interim device that, at least since the Industrial Revolution, is relied upon while awaiting a techno-logical fix.

Work has long been a basic feature of social planning all across the ideological spectrum. For, in a world in which at least some human beings need to work in order that their kind can survive and occasionally even prosper, it is difficult for a social planner not to take work into account. But in taking work into account, two different but inter-connected questions need to be addressed: (1) what work ought to be done; and (2) who (or what) ought to do it. The answer to the first question depends on goal-selection, for example, with regard to a desired level of need satisfaction and/or commitment to progress. The answer to the second at least presupposes a theory of distributive justice in society. More egalitarian ideals have been espoused in what I call the communitarian tradition.[2] But proponents of labor-saving devices are more likely to be associated with the authoritarian tradition.

Linked together as complementary themes of the authoritarian tradi-tion are (a) the claims of an elite to exemption from some or all work and (b) some rationale to justify imposing the burdens of work on a subservient class or classes of people. The principal reason given for requiring work of others has always been that they will not otherwise qualify for social benefits. This reason has, in turn, been mystified with reminders that people are being punished for something or other and/or are on probation for a less demanding life ahead. Typical reasons given for excusing oneself from work that one requires of others is that one has more (socially) important things to do or, simply, that one is "the boss."

Authoritarian strategy with regard to the workforce has long been characterized by an interim rule and an ultimate rule. The interim rule is: do what you must to get productivity out of your workers. The ultimate rule is: whenever possible, replace people with machines. The history of the interim rule reveals the origins of labor-saving in anti-worker bias. The history of the replacement rule shows directly the culmination of anti-worker bias in worker-displacing technology and indirectly the conceptual inadequacy of the technological fix.

The German sociologist Max Weber identified the making of money as the single-minded, joyless, and ultimately irrational goal of the work ethic.[3] And as he described it, this ethic could be interiorized only by the actual or would-be entrepreneur. There is, however, a corollary to the work ethic to the effect that poverty is not circumstantial but is a direct result of its victim's failure or refusal to work. In this latter form, the work ethic has been applied even to the unpropertied wage laborer as an incentive to greater productivity, by means as diverse as piece work and workhouses. As thus "applied," the work ethic has served as an interim strategy of the authoritarian philosophy of work. It is generally associated with the Protestant reformer John Calvin and the Calvinist Puritans. But it did not originate with Calvin; and its appeal transcends ideological boundaries.

Scholars following the lead of Marx and Weber have generally assumed that capitalism developed in the sixteenth century; that before that time society had been made up mainly of peasants for whom the principal economic unit was the extended family or, at most, the village; and that the destruction of the latter was more or less a *sine qua non* for the occurrence of the former. It is now recognized that socioeconomic realities were notably more complex than this, and that, in particular, the task of developing theological justifications of trade and commerce was already well underway in the thirteenth century. Up until that time, however, literate Christians in Europe tended to be as elitist and authoritarian on the subject of work as were their cultural forebears.[4]

Saint Benedict (d. 543) is retroactively credited with having made work and prayer (*ora et labora*) the dual objectives of the monastic order which he founded in Italy; and on this basis he is said to have restored to work the dignity that ancient civilizations had denied to it. This interpretation is, however, too simplistic. If anything, Benedict stands in history as a dutiful perpetuator of the values of authoritarian rule. He has no respect for "sarabites," because they are monks who consider whatever they think good or choose to be holy and whatever they do not wish to be unlawful. The life-style of "gyratory" monks, who are "always wandering and never stationary" (who, in a word, imitate in this respect Jesus and his apostles) is, in a word, "wretched." Only cenobites, who live together in total and unquestioning obedience to their abbot, are truly worthy monks. In his renowned and influential "Rule" he spells out just how the hours of the day are to be divided up, in the different seasons, between manual labor and sacred reading. The former is never to be so "violent" as to drive away the more feeble or delicate brothers; but, he also notes, "they are truly monks if they live by the labours of their hands; as did also our fathers and the apostles."[5]

Whatever Benedict's intentions, in the course of time this early commitment to manual labor was reduced to mere symbolism as the monks left hard work to serfs and wage earners and engaged themselves only in more honorific and less tiring endeavors such as baking, gardening, and brewing.[6]

The holy elitism that characterized "reformed" Benedictinism is also found among the (clerical) intellectuals of the high Middle Ages who depended on institutionalized thinking to earn their living. Thomas Aquinas, a thirteenth century Italian monk with aristocratic origins, repeated Benedict's endorsement of both contemplation and action, but was willing to separate one from the other on the basis of one's position in society. He agreed with Aristotle that contemplation is the highest human endeavor, to which the chosen few might devote all their attention. And he also agreed with St. Paul's admonition, "If any man would not work, neither should he eat" – but only to the extent that such work is necessary. Where it was "necessary" was among the common people, who accordingly had a right as well as a duty to work. They were not to question or attempt to rise beyond their "natural" station in life; but in turn they were entitled to a "just price" (*justum pretium*) for their labor, i.e., just enough to provide a bare livelihood for one's self and one's family.[7]

This essentially authoritarian view of appropriate compensation for work did accommodate a paternalistic exception for the disabled. Anyone who was truly unable to work was recognized as having for that very reason a right to be cared for. But it was no more obvious in the Middle Ages than it is today just who is sufficiently disabled to merit care without working. At the time of Thomas Aquinas there was, to speak anachronistically, a sizeable surplus pool of labor. And as a result begging was an accepted means of gaining one's livelihood – one to which the formerly wealthy St. Francis of Assisi gave spiritual dignity by requiring his monks to rely upon it for their daily bread.

This comparatively idyllic policy with regard to the work obligation came to an end in the first half of the fourteenth century, for reasons that are still not clearly understood. But just as commerce was beginning to expand in various directions, Western Europe was decimated first by famine and then by the Black Death, which in 1347 swept across the continent from Constantinople and by 1349 had eliminated over a third of the population of England. The resulting sharp reduction in the labor supply led to the Statute of Laborers (1349), which accomodated

English landowners' need for agricultural workers by forbidding the able-bodied to beg, travel, or demand more than customary wages and requiring them to labor for their livelihood. Subject to the penalty of imprisonment, unskilled people who survived the plague were thus circumscribed:

That every man and woman of our realm of England, of what condition he be, free or bond, able in body, and within the age of three-score years, not living in merchandize, nor exercising any craft, nor having of his own whereof he may live, nor his own land, about whose tillage he may occupy himself, and not serving another, if he in convenient service, his estate considered, be required to serve, he shall be bounden to serve him which so shall him require; and take only the wages, livery, meed, or salary, which were accustomed to be given in the places where he oweth to serve, the twentieth year of our reign of England, or five or six other common years next before.[8]

Or, as one modern commentator puts it, "The King and his lords saw begging, movement and vagrancy, and the labor shortage as essentially the same problem, to be dealt with in one law."[9] But the fabric of feudalism was coming undone, and the serfs' quest for freedom generated more and meaner laws. Persons without a letter authorizing travel were to be put in the stocks (12th Richard II, 1388). Yet after a century of thus battling against "idleness, mother and root of all vices," it was noted at the time of King Henry VIII that the number of vagabonds and beggars had actually increased. So provision was made for those truly in need (Statute of 1531); but able-bodied loafers were subject first to public whipping in the nude, then to whipping plus loss of part of one's right ear, then if still not willing to "put himself to labor like as a true man oweth to do," to "pains and execution of death" (Statute of 1536). During this same period of time most European countries empowered a new official, known in England as overseer of the poor, to put poor people to work and to imprison those who refused or performed unsatisfactorily (a somewhat primitive approach to vocational training). Meanwhile increasingly severe laws were being enacted to outlaw begging.[10]

It is in this context that Protestant reformers put forward their views about work. Martin Luther (1483–1546) still drew upon the just price theory to justify telling people to work at the trade or profession into which they were born. But he attributed equal value to any kind of work, active or contemplative, and stressed the religious dignity of one's work as a vocation or calling. Thus in his little book about vagabonds, *Liber Vagatorum*, he linked the Reformation to the growing movement against beggars by endorsing almsgiving only to duly certified indigents.

The Lutheran Eberlein proposed in his utopian *Wolfaria* (1521) to abolish serfdom, execute all mendicant friars, strictly regulate all trades and professions to avoid production of luxuries, and set everyone, including the nobility, to work at the only really honest occupation: agriculture. Johann Andreae, a Lutheran priest, dreamed of a society called *Christianapolis* in which perfect officials would not tolerate begging and would give material assitance to the poor only after careful examination of their needs.[11]

John Calvin (1509–1564) pushed the significance of one's work even farther by tying it in some inscrutable way to one's eternal salvation. So casual work is for this purpose inadequate, and dislike of work raises serious doubts about one's being among the elect. Although committed to a rigorous doctrine of divine predestination according to which human choice is irrelevant to the final outcome, Calvin insisted that the faith by which one is saved is expressed in and through methodical, disciplined, rational, uniform and hence specialized work. Puritanism, which was an offshoot of Calvin's teachings, drew the logical conclusion that wealth-seeking is a fine way to assure one's salvation; and in this way, according to Max Weber, Calvin's austere theology provided the ideological underpinning for capitalism. According to another interpretation, however, what Calvin provided was a religious justification for the capitalist's hard-nosed approach to discipline on the assembly line.[12]

The latter interpretation is certainly borne out by the technocratic moralizing of Scottish engineer Andrew Ure. Ure, the Calvinist ideologue of the Industrial Revolution, explained the need for hard work by associating its pain with that of the crucifixion of Christ.[13] But Calvinism was not alone in its endorsement of work either during the mass production or the mechanization phase of the Industrial Revolution. Robert Burton, a British don with no known religious preference, dreamed in his *Anatomy of Melancholy* (1621) of a society in which there would be no "beggars, rogues, vagabonds, or idle persons at all, that cannot give an account of their lives how they maintain themselves" and in which all able-bodied poor would be "enforced to work."[14] Two centuries later, Joseph Proudhon developed an anarchist glorification of labor around the idea that the value of work is directly proportional to how hard it is.[15] Karl Marx, by comparison, was not nearly so enamored with work, as will be noted below. But Andre Gorz, a contemporary Marxist, has said that "after the communist revolution we will work more, not less."[16]

Of course, these and other encomia of work, whatever their ideological roots, share a common flaw: even if a society depends on the work of some, that work will not be held in esteem, either by the workers or by their beneficiaries, if something else, notably ownership, is considered more honorific.[17] But the point here is simply that a work ethic serves the purpose of encouraging productivity in the absence of suitable machines. This is evident, for example, from the various workforce development proposals put forward in England during the sixteenth and seventeenth centuries. Faced with numerous and potentially revolutionary poor people, literate members of the leisure class came up with all sorts of ideas about how best to put the poor to work without upsetting the rich. Typically conservative ideas for social experiments of every sort, on both the community and the national scale, aimed at doing away with vagabonds and beggars, turning them into totally responsive instruments which, properly organized, would turn a profit for the enterprising rich.[18] This organization of "manu-facture," which culminated in Ambrose Crowley's authoritarian Law Books for his ironworks, won the praise of eighteenth-century landowner Adam Smith, who has immortalized "mass production" of pins. But even as this "assemblying" of workers became common, the spectre of "labor-saving devices" loomed on the horizon.

Gabriel Plattes argued that such labor-saving devices should not be adopted until a labor shortage has first developed. But John Bellers, a wealthy Quaker professionally concerned with poor relief, took the view that prohibiting a labor-saving device by law is like requiring a laborer to work with one hand tied behind his back. In spite of some legal approaches of the sort favored by Plattes, it was Bellers's openness to technology that served as a model for the entrepreneurs to come.[19]

III. MANAGERIAL RESPONSIBILITY FOR "LABOR-SAVING": THE ULTIMATE RULE

(3) Even if some people do intend to reduce the workforce by introducing a labor-saving device, they are not acting freely but only as they are compelled to do in response to competition.

Response: (a) Labor-saving devices are not necessarily adopted in response to competition; and (b) even if concern about competition is a consideration, a labor-saving device may not be a suitable response.

Already in the Middle Ages people took delight in various devices that could do something ordinarily done by humans. But in the absence of motivators like profit and progress, such devices were perceived merely as objects of wonder. As capitalism gradually transformed social goals, people learned to think of such devices as sources not just of wonder but of lower-cost productivity by means of which they might gain an advantage over their competitors. As we shall see, this is not necessarily so. Besides, labor-saving has not been espoused solely to beat the competition. It has also been motivated by a desire to separate workers from their jobs – either directly, as a result of a deliberate anti-worker bias, or indirectly, as a result of a pro-technology bias the consequences of which for workers are not well anticipated. The indirect bias will be considered in connection with the fourth counter-argument. Here only the direct bias will be considered, before addressing the relevance of competition.

A. The Direct Anti-Worker Bias

Almost from the onset of the Industrial Revolution, with its characteristic centralization of workers in capitalist-controlled plants, forces were set in motion that made the idea of production without payrolls attractive to the entrepreneurial class. Skilled workers who had previously enjoyed relative autonomy in their work life chafed at the impoverishing terms offered to them by the factory owners. The response of the owners, as often as not, was to look for a technological substitute for such intractable employees.

Thus, for example, did a certain Mr. Roberts respond to British textile entrepreneurs by developing a spinning automaton known as "the Iron Man" to displace high-wage skilled spinners. More generally, according to Andrew Ure,

Wherever a process requires peculiar dexterity and steadiness of hand, it is withdrawn as soon as possible from the *cunning* workman, who is prone to irregularities of many kinds, and it is placed in charge of a peculiar mechanism, so self-regulating, that a child may superintend it.

In this way, he notes, "when capital enlists science in her service, the refractory kind of labour will always be taught docility."[20]

A decision to replace skilled workers with some labor-saving device is seldom if ever going to be made, of course, just out of a desire for docility of the sort articulated by Ure. In the first place, science is not always prepared to provide industry with a quick technological fix, nor is labor saving the only motive for introducing new technology. A substantial part of good management strategy in this regard is to determine what is most advantageous in light of all known variables. Having determined this insofar as possible, one might in a given situation choose not to introduce an available device – until, that is, there is a significant change in one of the variables.

Standard dogma in this regard among neoclassical economists was to the effect that a rise in the cost of labor will precede a decision to mechanize. Introduction of "the Iron Man" into the textile industry is an example. Others are provided by Karl Marx. He notes how manufacturers in England turned to mechanization only after the Factory Laws limited child labor to four-to-six hour shifts and children's parents refused to sell their "half-timers" for less than full-timers. He also points to the practice of producing machines in one country to be used in another country where high wages motivate such substitution, and indicates that this very practice so expands the labor pool in the country where the machines are introduced that other industries there are spared the need to mechanize. For, says Marx, the capitalist's "profit comes . . . not from a diminution of the labour employed but of the labour paid for."[21] Thus even where displacement of workers by machines is prohibitively expensive if not still technically infeasible, the very threat of doing so might be employed to dissuade workers from demanding higher compensation.[22]

The need to make a profit is unquestionably an important reason for a company to look for ways to cut costs; but labor-saving is only one of the ways to cut costs, other possible ways being through cheaper raw materials or cheaper service on corporate debt. Besides, there are factors other than profit or revenue maximization that management might consider to be of overriding importance. For example, it is reported that a landowner in India will resist profit maximization of his agricultural business if by increasing his tenants' share he risks having his tenants pay off their debts to him and get out from under his control.[23]

That control over the workforce can even take precedence over profit maximization is central to the Marxist analysis of mechanization under capitalism. In Harry Braverman's view, for example, control of the work process is the overarching reason for mechanization. The "deskilling" of production that mechanization has effected is, according to Braverman, deliberate but not inevitable. In keeping with the Marxist class analysis of this process, he too speaks of capitalism striving for "domination of dead labor [machinery] over living labor [workers]," thereby attributing an anti-labor animus to the business decisions to mechanize and automate.[24] The evidence he is able to muster for this claim is largely circumstantial and anecdotal, for example, the fact that supervisor-operated numerical control (N/C) became the automation of choice rather than the equally efficient record-playback system (R/P) that leaves programming in the control of skilled machinists.[25] Historian David Noble has since elaborated upon this example in impressive fashion, but without finding a "smoking gun."[26] Other Braverman inspired studies of the labor process also lend support to the claim that skilled industrial workers are being systematically displaced by machinery.[27]

Whether and in what numbers workers affected by deskilling ever rejoin the workforce is a subject of much debate. What seems clearly beyond debate is that the immediate purpose of a labor-saving device is, as its very name indicates, to save labor, i.e., to save a company some of the costs associated with paying labor. This means that the process of introducing such a device is inherently even if not intentionally hostile to anyone whose labor is thereby going to be "saved." Marx and many who have followed him would go on to argue that there is a definite and deliberate anti-labor bias on the part of management that tilts cost-cutting decisions whenever possible in the direction of cutting payrolls.

A generalized claim to this effect is difficult if not impossible to prove. (Braverman says this is because the complexity of reasons why there has been a "transformation of the labor process" does not lend itself to a "unitary answer.")[28] But many on the side of management, especially industrial design engineers, have displayed just this sort of anti-labor bias. Ure, for example, put it thus:

It is, in fact, the constant aim and tendency of every improvement in machinery to supersede human labor altogether, or to diminish its cost, by substituting the industry of women and children for that of men; or that of ordinary labourers for trained artisans.[29]

The century and a half since Ure has been characterized by a variety of strategies to hold down the cost of labor in manufacturing, including the total regimentation of company towns and Frederick Taylor's preference for "stupid" workers who can be counted on to be docile, advancing to various managerial theories of "job enrichment" to pacify the ever more demanding workers of the twentieth century, and culminating in an all-out effort to design workers right out of production processes entirely.[30]

As this trend clearly indicates, technological unemployment is inevitable, to a degree still subject to debate. It is inevitable, however, not by virtue of any law of nature or because there are no alternatives with different consequences but because robots and other microelectronic devices are already perceived as cost effective in the long run and hence a necessary condition for staying competitive in the industries affected.[31] Thus is being carried out Ure's "automatic plan," which he defined as follows: "(S)killed labour gets progressively superseded, and will, eventually, be replaced by mere overlookers of machines."[32] And to this day engineers seem to believe that automation can be cost effective only by eliminating humans, because the greatest expense is incurred in trying to accommodate "man in the loop." Says Lawrence B. Evans, an MIT chemical engineer:

The cost of complex electronic circuitry continues to decrease exponentially (by a factor of about 1/2 each year) due to large-scale integration (LSI) semiconductor technology. . . . The real cost of a system is in the hardware for communication between man and that system (displays, keys, typewriters) and this cost is a function of the way the system is packaged. Thus, automation functions and data processing become economic if they can be done blindly, without the need for human communication.[33]

Estimates vary as to just how much less expensive it may be to use robots in place of humans; but that there will be significant savings is widely assumed. As one writer puts it, a Japanese robot in automotive production can do at $5.50/hour what a UAW worker does for $18.10/hour (wages and fringe benefits).[34] An estimate of this sort is typically based on a comparison between costs incurred from labor and costs of procuring and maintaining a robot. Robot providers claim that robot costs will be recouped within a three-year payback period from savings in labor alone. Of course, assumptions with regard to the cost of money, the cost of installation, and the cost of power and maintenance of a robot need to be adjusted up to take inflation into account. But the initial cost of producing a robot may well drop from, say, $50,000 in 1980 to just $10,000 in 1990. So recent estimates are probably at least in the correct order of magnitude.

B. Technology and Competition

The foregoing are clearly indications of free choice in the decision to adopt a labor-saving device. But, it may objected, they are not persuasive because they ignore the background conditions that mandate such a decision in the first place. These conditions, it is argued, involve competition among firms each of which seeks to control or at least to constrain others. It does not follow from this, however, that a labor-saving device is the appropriate response.

The labor-saving device has acquired a reputation for being able to deliver a technological fix to any business or industry that has what is considered to be an excessive payroll. This reputation, in turn, enhances significantly the success rate of suppliers who want to sell labor-saving devices to corporate buyers. What, if anything, the latter or anyone else stand to gain from any such purchase is, however, by no means obvious.

That society may wind up worse off because of corporate commitments to automation may be seen either on the level of the affected individuals or on the level of society taken collectively. It is beyond dispute that individual workers are harmed by automation and that society is seldom able to undo all the harm. And as for society, the automation decisions of corporations are not made with a view to benefitting society as a whole. With a view rather to making money for their investors, companies being tempted by automation take only their own ("internal") costs into account, not the overall ("external") costs, direct and indirect, that spill over onto society in the wake of major technological change. The overall costs, most of which the comparatively defenseless members of society are usually called on to bear, can be traced for the most part to the loss of jobs in a society that distributes benefits on the basis of one's employment.

What is especially deserving of our attention is the question of whether, or under what circumstances, technology in the guise of a labor-saving device is likely to benefit its corporate buyer. One might suppose in this regard that any corporation that could consistently answer these questions would know how to allocate R&D money for the purpose. But the matter is not that simple, in part because of uncertainty about the performance of the device and in part because of uncertainty about the performance of competitors who are also in a position to adopt a functionally comparable device.

Why there might be uncertainty about the performance of a device is abundantly illustrated in the history of technology, for example, the development of machines to do calculations, from the counting board to the computer.[35] Inventors such as Leibniz, Pascal, and Babbage produced mechanical devices that could do calculations accurately, but not rapidly enough to make it economically attractive to substitute them for human clerks and accountants. Electricity introduced a new factor into the equation, but further development had to await improved designs as well as a situation of cost-discounting urgency. The Allies in World War II needed to expedite the development of trajectory charts for new types of shells, which were being introduced into the arsenal faster than human calculators could keep up. The needs of the war effort outweighed expenses as the ingenious ideas of Vannevar Bush and others were applied to produce the first workable computer. That it was able to perform its tasks as expected is, however, irrelevant to considerations of economy. There simply was no upper limit of expense beyond which R&D would not go under wartime conditions. This situation was not without precedent, of course, and it continues to this day – not with a view to winning a war, necessarily, only with a view to "defense."

Economically unjustified introduction of labor-saving devices does not occur with such indifference to cost in the private sector. But this does not preclude a corporate commitment to high-risk adventure, for reasons having to do with everything from the corporate image to the corporate will to dominate an industry. The former is often a shortcut to bankruptcy; the latter, a strategy best left to oligopolists. Thus, for example, might narrow-gauge cost considerations about the introduction of robots give way on occasion to a desire for product quality improvement, for example, in production of Chrysler's K-car at the Newark, Delaware, plant and of GM's Fleetwood in Detroit, where $8.5 million of robots save only $120,000/year. In such instances, a more affluent market is targeted, and cost is expected to be recouped through sales.[36]

It is in this context that one needs to consider neoclassical economic theories about motivation for technical change.[37] On the assumption that the entrepreneur rationally selects that mix of labor and capital that will maximize net revenue or profit, the orthodox position before World War II was that labor-saving innovations come about as a direct result of high-priced labor (and similarly with regard to the price of capital). The logic in that position is faulty, since total costs of production can be brought down by reducing the cost of either labor or capital. So there would be no special reason to concentrate on reducing the cost of labor unless one has either easier access to labor-saving knowledge or, as in the case of Ure, an anti-labor(er) bias.

The orthodox position is even less persuasive if confronted with game theory considerations about the disparity between collective and individual benefits to be derived from a labor-saving innovation. If it be assumed that no entrepreneur knows what any other is going to do, each is caught in a Prisoners' Dilemma over how to avoid being the only one to bear the cost of introducing, or of not introducing, the labor-saving innovation. If, in addition, it be assumed that a search for cost-reducing innovations will be the same whether one focuses on saving labor or on saving capital, then in the absence of some special reason to favor labor saving, for example, a history of rising labor costs, the choice of direction might depend on nothing more than a flip of a coin.

Suppose, finally, that, an entrepreneur knows that one or more competing entrepreneurs might invest in a labor-saving device. *Alteris paribus*, it is still not necessarily the case that the entrepreneur armed with this information should invest in the same labor-saving device. For, if one's competitors do in fact "save labor" with their new device, the cost of labor in the industry will fall, and our non-innovating entrepreneur will be able to undercut the competition by virtue of both having avoided the cost of innovation and benefitting from the availability of cheaper labor. Since, however, the other entrepreneurs can be presumed to be making the same assessment, the entrepreneur must also contend with the possibility that none will innovate, in which case the cost of labor will continue to rise and the advantages of innovating unilaterally will increase. In short, introducing the labor-saving device would be the rational profit-maximization course of action only for the entrepreneur who does so unilaterally, not for those who "follow the crowd." (What better counter-argument to the old saw: Be not the first by whom the new is tried nor yet the last to cast the old aside?)

The resulting quandary about whether to invest or not in the labor-saving device has been described as a game of Chicken. It is such, however, only if (a) the innovation in question merely substitutes for rather than improves upon the product and/or its production and (b) no competitor does in fact make a move to introduce the innovation. Practice with regard to introducing innovations into traditionally labor-intensive industries in the 1980s satisfies, or at least is perceived as satisfying, both of these provisos. So in the real world that theoretical quests for perfect rationality have so far failed to model, the game being played is not Chicken, but Corporate Survival; and the only thing certain is that the workforce will not be the winner.

IV. THE EUDAEMONIC RATIONALE

(4) Even if some human beings do intend the unemployment consequences of labor-saving, this does not prove that they are acting against the interests of workers. For, the long term result of this process will be a state of affairs in which people will no longer need to work.

Response: The end envisioned is not easily extrapolated from available data; and, this notwithstanding, felicitous ends do not justify unjust means.

Working class endorsement of the saving of labor is not a spontaneous tropism on the part of those who live too close to labor. It has been instilled over the centuries by promises of a work-free utopia to come. Current versions of these hoary promises go beyond their classic predecessors by noting a need to re-educate people accustomed to labor for a life of leisure. There are basically just two versions of the utopian scenario. One recognizes a painful transition as we pass through the tunnel on our way to the light. The other, notably more pollyannaish, skips over interim *angst* to concentrate on the latter days.

The rationale here runs more or less as follows. For the first time in history technology has brought us to a point at which there are not and will never again be enough jobs to go around. So, whatever may have been society's need for workers in the past, that need is with us no longer. We have reached a point which has all along been proclaimed as a goal of technological innovation, namely, to eliminate the need for "labor" (meaning paid workers) by utilizing machines instead. Only so long as technology could not deliver on this promise, the obsolescence argument would continue, was there any need to instill in *humans* a sense of duty with regard to work. Now this need is passing from the scene, so society must transform its values and its objectives accordingly.

The pollyannaish proposal, presented as a prediction, is that society must begin to focus not on work, which is no longer an appropriate goal, but on leisure, which we shall have in the future whether we are ready for it or not. In other words, work is less valuable now than it once was because it is less important to the satisfaction of human needs than it used to be. Robert Theobald, for example, envisions the emerging situation in this way:

[In t]he new society we are entering . . . [m]any people will tend to work intensively for a period of time and then need to re-create and re-educate themselves. We shall extend rapidly the concept of sabbaticals in terms of the number of people involved, the number of occupations for which they are considered relevant, and the length of time for which people can free themselves up from responsibilities. Societies will be able to make free time available because of the impacts of computers and robots, which will limit the amount of human energy needed for industrial-era jobs.[38]

None of this, by the way, would have saddened Marx and his followers, provided only that control of the means of production be not in the hands of the capitalists but of the proletariat as represented by the State. Marx in his youth envisioned a future in which machines would be doing the work and the only question would be who was going to reap the benefits. He did not, to be sure, think of the road to such a state of affairs as one to be easily traversed. Respectfully rejecting the escapist proposals of the utopian socialists, both Marx and Engels sought to confront the new industrial reality head-on to salvage a future for the vast majority of human beings who make up the working class. On this view, labor, however valuable in the capitalist setting, has no intrinsic value. Machines are welcome as a means to the eventual liberation of human beings from dehumanizing drudgery. Lenin in the interim welcomed even Taylorism as an appropriate device for increasing productivity.[39] Workers are alienated from their work in the typical factory system, but this is due not to rationalization of work but to capitalist ownership. Scientific socialism promises the surmounting of alienation, first by assuring workers that collectively they own the means of production, and in time perhaps by freeing them of responsibility for production and handing this over to machines. Marxism, then, does not romanticize work but rather socializes the work ethic for the duration of our dependence on human labor for productivity.

In this regard, Harry Braverman acknowledges that no less a deskilling mechanization has taken place in Communist countries; but he excuses deskilling there because it is merely imitative of what capitalists did and expresses the hope that in these countries the dominance of machines over people is only transitional.[40] An even more utopian expression of hope in this regard is that of another neo-Marxist, Herbert

Marcuse, who welcomes automation in spite of the short-range concerns of workers. These concerns, says Marcuse, are legitimate in the absence of "compensating employment." But, he insists, over the mid-range of time such opposition to technical progress prevents "more efficient utilization of capital," "hampers intensified efforts to raise the productivity of labor," and leads to economic crisis and exacerbation of class conflicts. That is bad enough. What is worse is that opposition to automation stands in the way of eventual attainment of a liberating utopia based on technology. In Marcuse's words:

Complete automation in the realm of necessity would open the dimension of free time as the one in which man's private *and* societal existence would constitute itself. This would be the historical transcendence toward a new civilization.[41]

Neo-Marxist Andre Gorz is also convinced that technology has prepared "paths to paradise." He is careful to point out, however, that these paths will be traversed toward "liberation from work" only "within a social environment which does not yet exist (at least not generally)."[42]

Whether or not the ultimate outcome of all of this will be anything like the downfall of capitalism that Marx predicted is a matter for specialists to debate. And as they do so they will want to bear in mind that the assumptions of classical economics, against the background of which Marx developed his alternative theory, are inadequate to represent the complexities of today's transnational marketplace of competing forces. What is important in the present context, somewhat more simply, is the impact that the delaboring of productivity will have on people regardless of the sociopolitical framework within which their onetime workplaces happen to have been located. On this issue divergent ideologies are of only secondary importance. More basic is the cross-cultural heritage of the human race with regard to the value and necessity of work, which finds its way into the views of writers of every persuasion.

Take, for example, Herbert Marcuse's guarded optimism about the benefits that labor-saving technology will bestow upon the working class. That Marcuse placed so much trust in the eventual blessings of technology for the working class reflects his interest in Marx's earlier writings.[43] But it does not reflect his interest in the writings of Freud. Although Freud seldom addressed the subject of work directly, one footnote in his *Civilization and Its Discontents* states explicitly what elsewhere is only implicit. Work, he says here, is the best means of tying an individual to the community and, if it is work at a profession, is an excellent instrument of sublimation. But, he regrets,

as a path to happiness, work is not highly prized by men. They do not strive after it as they do after other possibilities of satisfaction. The great majority of people only work under the stress of necessity, and this natural human aversion to work raises most difficult social problems.[44]

Thus is suggested the view that work, even if eventually unnecessary for economic productivity, is nonetheless an important vehicle of human creativity. This view underlies the concerns of Erich Fromm about the possible demise of work.[45] And a century earlier it underlay William Morris's aesthetic of work and Proudhon's glorification of work as having intrinsic dignity.[46] It was important to William Wordsworth and to John Ruskin as they watched cottage industries giving way to dehumanizing division of labor in factories. And in our own day it is finding expression in the works of novelist and poet Marge Piercy.

Emphasis on the human need for work is a key feature of E. F. Schumacher's insistence that we move toward "appropriate technology." As he once expressed his ideological assumption, at least for the poor man, "the chance to work is the greatest of all needs, and even poorly paid and relatively unproductive work is better than idleness."[47] Schumacher himself tried to incorporate this pro-work view into a kind of Buddhist economics that stresses the importance of work to the individual and to the community. But it would not be difficult for the authoritarian tradition to co-opt such humane theorizing for ends quite unrelated to anything that Schumacher wanted for the world. It was, after all, just this sort of co-opting that lay at the foundations of the modern factory system. The disciplined religious community that Saint Benedict developed in the sixth century became a thousand years later a model for organized "manufacture" which was well in place long before the appearance of mechanized assembly.

These worries aside, there is still good reason to support the claim that work, even laborious work, is a valuable instrument of human fulfillment. This admittedly has not been proved to be true universally and without qualification. And, like claims with regard to our reliance on gravity before humans experienced weightlessness, its truth may turn out to be of somewhat limited applicability. But just to the extent that it still obtains in the world as we know it, the delaboring of production should not be endorsed without serious qualification. In particular, it should not be endorsed so long as work remains the principal means of support for oneself and for one's dependents while at the same time welfare rights are viewed not as the fruit of human progress but only as a recipe for sloth.[48]

This sort of cultural Catch-22 is not inherent in our genes. It is, if you will, the result of holding on to the interim rule even though the ultimate rule is now fully operative. Of far greater significance, however, is the fact that human beings – and, not infrequently, identifiable human beings – are responsible for the perpetuation of these rules. For, this constitutes a *prima facie* case of responsibility for the consequences of their application. A defense based on "business necessity" could no doubt be mounted. But, I submit, this defense can and should be overcome by showing that there are alternative social solutions.

Indiana University/Purdue University at Indianapolis

NOTES

[1] The hard determinism implicit in the view of Jacques Ellul is open to an unusually subtle qualification. Only his "sociological" works require a determinist conclusion; his theological works offer an escape in Christian activism. See my review in *Nature and System* 3 (September 1981): 184–188.

[2] The communitarian tradition, especially as exemplified in utopian literature, looks to reform and reconstitution of social organization as the proper remedy for work-related inequities. See Frank E. Manuel and Fritzie P. Manuel, *Utopian Thought in the Western World* (Cambridge, MA: Belknap/Harvard University Press, 1979).

[3] *The Protestant Ethic and the Spirit of Capitalism*, trans. Talcott Parsons (New York: Scribner's, 1958; German original, 1904–1905), p. 53.

[4] See Jacques LeGoff, *Time, Work, and Culture in the Middle Ages*, trans. A. Goldhammer (Chicago: University of Chicago Press, 1982); Alan Macfarlane, *The Origins of English Individualism* (New York: Cambridge University Press, 1978).

[5] The Rule of St. Benedict, in *Select Historical Documents of the Middle Ages*, trans. and ed., E. F. Henderson (London: G. Bell and Sons, 1925), pp. 274–275ff.

[6] LeGoff, *op. cit.*, p. 84; James W. Thompson, *The Economic and Social History of the Middle Ages (300–1300)* (New York: Century, 1928), pp. 607–618.

[7] Adriano Tilgher, *Work*, trans. D. C. Fisher (New York: Arno Press, 1977; original, 1930), pp. 39–42.

[8] 23 Edward III, Statute of Laborers, 1349, quoted by Karl de Schweinitz, *England's Road to Social Security, 1349 to 1947* (Philadelphia and London: University of Pennsylvania and Oxford University Presses, 1947), p. 6. See Henderson, *op. cit.*, pp. 165–168.

[9] De Schweinitz, *loc. cit.*

[10] De Schweinitz, *op. cit.*, pp. 20–38.

[11] J. C. Davis, *Utopia and the Ideal Society: A Study of English Utopian Writing 1516–1700* (Cambridge: Cambridge University Press, 1983, 1981), p. 72.

[12] Tilgher, *op. cit.*, pp. 59–60.

[13] Andrew Ure, *The Philosophy of Manufactures or an Exposition of the Scientific, Moral and Commercial Economy of the Factory System of Great Britain* (London: C. Knight, 1835), p. 423.

[14] Davis, *op. cit.*, p. 100.

[15] Manuel and Manuel, *op. cit.*, p. 771.

[16] 'The Tyranny of the Factory: Today and Tomorrow', in *The Division of Labor*, ed. Andre Gorz (Sussex, England: Harvester, 1976), p. 58. More recently, Gorz has bought into the dream of a technologically generated utopia. See below in connection with note 42.

[17] See Thorstein Veblen, *The Theory of the Leisure Class* (1899) (New York: Modern Library, 1934; original, 1899), pp. 92–95, 231.

[18] Davis, *op. cit.*, chap. 11, pp. 299–367.

[19] *Ibid.*, pp. 318, 345–346.

[20] Andrew Ure, *Philosophy of Manufactures* (3rd ed.; New York: Burt Franklin, 1969; this ed., originally, 1861), pp. 19, 366–368. See also pp. 16, 40–41, 331, 369–370.

[21] Karl Marx, *The Poverty of Philosophy* (1847), in Marx and Engels, *Collected Works* (London: Lawrence and Wishart, [nd], vol. 7, pp. 207, 393–394. See Jon Elster, *Explaining Technical Change* (Cambridge: Cambridge University Press, 1983), pp. 163–171.

[22] David Dickson, *The Politics of Alternative Technology* (New York: Universe, 1975), pp. 72–73, 181–182.

[23] A. Bhaduri, 'A Study in Agricultural Backwardness under Semi-Feudalism', *Economic Journal* 83 (1973): 120–137. See also S. Marglin, 'What Do Bosses Do?', in Gorz, ed., *op. cit.*

[24] Harry Braverman, *Labor and Monopoly Capital* (New York and London: Monthly Review Press, 1974), pp. 193–194, 199, 227–228. See also p. 188.

[25] *Ibid.*, pp. 196–206.

[26] David F. Noble, *Forces of Production: A Social History of Industrial Automation* (New York: Knopf, 1984). See also James Fallows, 'A Parable of Automation', *New York Review of Books* 31, no. 14, September 27, 1984, pp. 11–17.

[27] See Andrew Zimbalist, ed., *Case Studies on the Labor Process* (New York and London: Monthly Review Press, 1979); Dan Clawson, *Bureaucracy and the Labor Process: The Transformation of U.S. Industry, 1860–1920* (New York and London: Monthly Review Press, 1980). Also instructive in this regard is the earlier work of George E. Barnett, *Chapters on Machinery and Labor* (Carbondale/Edwardsville: Southern Illinois University Press; London/Amsterdam: Feffer & Simons, 1969; original, 1926).

[28] *Ibid.*, p. 169.

[29] Ure, *op. cit.*, p. 23.

[30] See Edmund F. Byrne, 'Robots and the Future of Work', in H. Didsbury, ed., *The World of Work* (Bethesda, MD: World Future Society, 1983), pp. 30–38.

[31] Dickson, *op. cit.*, pp. 72–73, 181–182. Compare Ure's views on the need for machines to outdo foreign competition, *op. cit.*, pp. 31–32, 329.

[32] Ure, *op. cit.*, p. 20. See also pp. 1, 20–21, 23.

[33] 'Industrial Uses of the Microprocessor', in T. Forester, ed., *The Microelectronics*

Revolution (Oxford: Basil Blackwell, 1980), p. 144; originally published in *Science*, 18 March 1977.

[34] E. Janicki, 'Is There a Robot in Your Future?' *The Indianapolis Star Magazine*, November 22, 1981, p. 55.

[35] See Arthur W. Burks and Alice R. Burks, 'The ENIAC: First General-Purpose Electronic Computer', *Annals of the History of Computing* 3 (October 1981): 332–336, 386–388.

[36] *The Impacts of Robotics on the Workforce and Workplace* (Pittsburgh, PA: Carnegie-Mellon University, June 14, 1981), p. 52.

[37] The following analysis is derived from Elster, *op. cit.*, pp. 96–111.

[38] 'Toward Full Unemployment', in H. Didsbury, ed., *The World of Work* (Bethesda, MD: World Future Society, 1983), p. 54. For Frithjof Bergmann's view, see, for example, 'The Future of Work', *Praxis International* 3 (October 1983): 308–323.

[39] Dickson, *op. cit.*, pp. 55–56 and, in general, 41–62. See also Bernard Gendron, *Technology and the Human Condition* (New York: St. Martin's, 1977).

[40] Braverman, *op. cit.*, pp. 15–16, 22, 24.

[41] *One-Dimensional Man* (Boston: Beacon, 1966), pp. 35–37. See also pp. 44–45, 59, 231–232, 235. Marcuse bases his view on a passage from Marx's *Grundrisse der Kritik der politischen Oekonomie* in which Marx declares that labor time will eventually cease to be the measure of wealth.

[42] *Paths to Paradise: On the Liberation from Work*, trans. M. Imrie (London and Sydney: Pluto Press, 1985).

[43] Adam Schaff explains Marx's early dream of an "end of labor" as "youthful folly" categorically rejected in *Capital*. According to Schaff, "utopian prophecies" about what automation might accomplish "do not take us a single step further in the organization of our life today"; see *Marxism and the Human Individual*, trans. Olgierd Wojtasiewicz, ed. Robert S. Cohen (New York: McGraw-Hill, 1970), pp. 124–126, 134–135.

[44] *Civilization and Its Discontents*, trans. James Strachey (New York: Norton, 1962; original, 1930), p. 2, note. See Georges Friedmann, *The Anatomy of Work*, trans. Wyatt Rawson (New York: Free Press, 1964), p. 126; Philip Rieff, *Freud: The Mind of the Moralist* (3rd ed.; Chicago: University of Chicago Press, 1979), p. 245.

[45] *The Sane Society* (New York: Rinehart, 1955), pp. 288–289, quoted in Friedmann, *op. cit.*, pp. 45–55.

[46] Manuel and Manuel, *op. cit.*, pp. 745–747, 769. Note in particular the authors' comments about the nineteenth-century debate regarding the value of work, p. 745.

[47] 'Social and Economic Problems Calling for the Development of Intermediate Technology', mimeographed (undated), quoted in Dickson, *op. cit.*, p. 153.

[48] See David Macarov, *Work and Welfare: The Unholy Alliance* (Beverly Hills: Sage, 1980); Frances Fox Piven and Richard A. Cloward, *Regulating the Poor: The Functions of Public Welfare* (New York: Vintage, 1972).

STANLEY R. CARPENTER

SYMPOSIUM ON APPROPRIATE TECHNOLOGY

I. A Conversation concerning Technology: The "Appropriate"
Technology Movement

INTRODUCTION

In this paper I propose to discuss the topic of technology from a praxical perspective. The aim is to accomplish two objectives: first, to assess the residual effects of the social movement of the Seventies called "appropriate technology"[1]; and secondly, in the course of this assessment, to defend a mode of philosophical discourse concerning technology that is anti-foundational and historicist.[2]

By abandoning an essentialist or realist epistemological approach I may seem to be giving up the possibility of uniquely distinguishing the philosopher's role in discussions of technology from that of other cultural actors: scientists, engineers, politicians, feminists, ecologists, historians, dramatists, poets, etc. This surmise is well founded. That is to say, the project, so clearly articulated by Descartes and Locke and given definitive formulation by Kant, may be seen as problematic in light of current philosophical discussions.[3]

Denying, however, that philosophy can function as ultimate epistemological arbiter of every system of knowledge claims does not mean silence. A humane exercise of the skills of philosophical argumentation can contribute to pluralistic conversations about timely and controversial issues of the day. Focusing on the "appropriate technology" (AT) movement may, however, seem passé. As Langdon Winner notes, the AT movement lasted about four years and came to an end with Ronald Reagan's first inauguration.[4] Why then bother about AT? AT claimed to say something important about the social properties of technological devices. It espoused values that were widely admired such as promotion of resource conservation and sustainable technological processes, along with an emphasis upon greater democratic control of technological processes.[5] It elicited significant creative energy, especially from young people, and provided a positive alternative to the negativity of the Viet Nam protests. Its demise, however, has left a small number of lasting effects. Winner locates this legacy in the realm of ideas: a broader interpretation, for example, of such terms as "efficiency," "productivity,"

87

Paul T. Durbin (ed.), Technology and Contemporary Life, 87–105.
© *1988 by D. Reidel Publishing Company.*

"rationality," beyond their standard economic usage.[6] This would be a good thing, however modest.

More often than not, however, discussions about AT failed to change anything. It was frequently the case that arguments offered *pro* and *con* did not engage the other side at all. As a sometime participant in these discussions, I have myself been puzzled at the ineffectual results of the AT movement. It is as if the criticisms offered by proponents of AT about conventional technological practices had no bite; they failed to challenge the opposition because they were perceived as irrelevant, innocuous, or sadly lacking in a mature appreciation of the "facts of life" regarding the production system. One suspects that there were and remain alternative "rhetorics of technological possibility" at work. That is to say, the very conception of "feasible" options was not shared by proponents of AT and technicians of the conventional technical order.

Because, as Winner observes,[7] proponents of AT were generally ignorant of the history of technology, they naively believed that cataloging the faults of modern industrial technology, would lead naturally to positive reforms. On the other hand, because the defenders of the present order were themselves so steeped in a system of two-hundred-year-old assumptions – which tied together technology and a particular mode of production – as to severely limit their horizons of technical possibility, their attitude toward AT was either to ignore it or, in the best of cases to be discussed below, to grant certain of its reforms at the margins of current practice.

The order of presentation I shall adopt below is as follows. I shall first describe and defend an approach to the philosophy of technology that is anti-foundationalist. It will be located within broader contemporary epistemological arguments *pro* and *con* as to the nature of philosophical discourse itself. At issue in particular is the cogency of the view of philosophy as that which grounds and delimits human self-definition. In rejecting this point of view, I shall describe what role philosophy *can* productively play, however provisional and historicist its practice may be.

Adopting this stance, I shall look at a critique of AT by Harvey Brooks. It will serve to introduce salient features of the AT program but will itself be criticized for its unreflexive character. Categories of analysis which could lead to "creative redefinitions" of technology in society will next be outlined. I will propose that future efforts to define and devise appropriate technologies will be more successful than the AT movement discussed here if they involve conversations about technology that more completely recapture the social and political implications of proposed alternative embodiments. Philosophers of technology can assist in these discussions by showing how today's "common sense" about desirable configurations of technology was yesterday's debatable perspective. It may thus be possible to reinvite consideration of the "road not taken."

"GROUNDING" THE CONCEPT OF "APPROPRIATE TECHNOLOGY"

Some philosophers are fond of "grounding" beliefs, concepts, cognitive processes, morals, etc. Grounding is sometimes thought to be the definitive and unique contribution of philosophy to culture. Grounded concepts possess a certainty, a necessity, that can rout skepticism by prescribing canonical forms of rational discourse. Whereas the Greeks saw this grounding as coming from the nature of things, Kant argued that it reflected the nature of our minds. Whereas modern empiricists such as Bertrand Russell detected the origins of knowledge in the immediacy of sense data and in univocal logical necessity, some modern philosophers locate it in the necessary structures of language.

Particularly during the first half of this century, experimental science was viewed as the paradigm of rationality itself. Philosophy turned to the grounding of science in an attempt to articulate a canonical structure of theories and hypotheses along with their essential connection to facts. This, it could be argued, was but the latest of Western philosophical attempts to escape history, culture, and the vagaries of transitory natural languages. By holding actual scientific practices to the ideal yardstick of an organon of the scientific method, our certainty concerning the nature of things was progressively enhanced. Refined theories approached "nature's laws." "True" theory could expose the unbreachable limits of technological possibility.

As a subcase, the understanding of human nature likewise took a scientific turn. While it is true that the physical/mental dichotomy remained to plague behavioristic models of human nature, a description of human physiology, along with characteristic models of human behavior, was believed to be scientifically ascertainable, at least in principle. With the essential properties of human behavior identified, it should be possible, again, in principle, to detect and order human preferences. The combination of what is scientifically known about the physical world with what is similarly known about human preference behaviors, would make it possible to arrive at a fundamental description of appropriate technological action. An appropriate technology could thus be considered to be one which facilitated an optimal realization of preferences.[8]

Foundational epistemological programs, of the type just outlined, have come under attack in recent times. The grounding of a philosophy of technology along realist lines would require clear differentiation between the conceptual and the empirical, the intentional and the extensional, or in philosophical jargon, the analytic *a priori* and the synthetic *a posteriori*. It is these dichotomies that were significantly challenged by W. V. Quine in his much-discussed paper, "Two Dogmas of Empiricism."[9] Here Quine questions a Russellian empiricism (or the Wittgenstein of the *Tractatus*) according to which concepts *necessary* for constituting experience can be categorically distinguished from particular reports of experience. By extended and subtle argument, focusing in part on familiar difficulties encountered in translating from one natural language to another – from, say, the native tongue of the anthropologist to that of a foreign culture being observed – Quine disparages attempts to differentiate sharply between a foreigner's definitions, embedded logical structures, and patterns of reasoning, and other statements that are merely reports of experience. He suggests that the anthropologist's difficulty in ascertaining which statements to accept at face value and which to ask for additional support is not necessarily a reflection of a lack of knowledge of the alien culture, but, rather of the untenability of the language/fact distinction itself. Reflexively, he extends the point to the scientist's own language and thus calls into question categorical distinctions between questions of meaning and questions of fact.

The effect of Quine's pragmatic line, as well as that of contemporary Continental deconstructionist thought, is to regard language use more as a "writing" of facts than a factual "reading out" from some neutral material matrix. Criteria of truth have thus to do with degrees of coherence among our attempts to make sense of experience rather than correspondences between our readings and something "out there" which "true" readings may be said to reflect.

Thomas Kuhn's *The Structure of Scientific Revolutions*[10] represents another anti-realist attack on essentialist epistemology. By questioning the positivistic demarcation of the context of justification (a topic for philosophical analysis), from the context of discovery (of interest to psychologists, historians, sociologists, etc.), Kuhn distinguishes what philosophers do from what everyone else does as matters of degree rather than kind. Paradigms, like games of all sorts, possess great utility. They define limits, create selective foci of interest, channel energies, and erect standards of proof. That philosophers are more comfortable analyzing the structures of theories and the standards of proof that characterize a particular paradigm than they are recording meter readings speaks to their "logophilia" rather than their disciplinary uniqueness. A holistic picture of science, such as Kuhn provides, sees no niche to be occupied by the philosopher as bearer of a "meta-paradigmatic" vision of the "scientific method." There are, rather, many methods, many exemplars, and many disciplinary matrices,[11] so that for the scientifically inclined philosopher, with that characteristic philosophical propensity for generalization and the fostering of systematic coherence, there is a position on the field to be filled as a team player but not as a referee.

This line of argument would call into question any attempt to explicate the phrase "appropriate technology" along essentialistic lines. Upon rejecting the idea of a sharp philosophy/science distinction, as well as the claim that knowledge is culturally dependent and linguistically idiosyncratic, we ought rather to expect any definition of technology to bear marks of its cultural ancestry. Our conception of technology, and by extension our ideas about appropriate technology, is thus bound to be situational, reflecting changing human interests, intentions, and aspirations. It is, in fact, a cultural artifact of our time. Margolis calls this approach "praxical."

This indissoluble linkage between the purposive activity and work of man and what may be called the legibility of nature (including, a fortiori, human culture and history) is, on the least tendentious reading, what is meant by the *praxical*. . . . [A]dmitting the praxical does entail (even in its most neutral form) rejecting all forms of naive or direct realism – in effect, all forms of the cognitive transparency of nature, in particular, of essentialist interpretations of the laws of nature. The reason is simply that, on the admission, *our very understanding of the world is seen to be a function of the contingent, varied, and clearly transient forms of human intervention.*[12]

It is easy enough to overlook the transitory and contingent character of modern technology, of which Margolis here writes. In the modern industrial order we are presented, in Winner's apt phraseology, with a world "of material accomplishments and social adaptations of astounding completeness."[13] The fact is that contemporary life is embedded in an artificial environment. It is much easier to lapse into an uncritical technological determinism – this is the way things have been since the Industrial Revolution occurred – than to make the effort to identify alternative technological solutions. Furthermore, the task of identifying alternatives is simplistically approached when the social and political features of the alternatives are discounted or not perceived. Such a criticism can be applied not only to the defenders of the technological *status quo* but also to "hardware-oriented" proponents of AT.

For the purpose of introducing the main features of AT, I would now like to examine a critique of AT provided by Harvey Brooks.[14] His treatment is informative and, within the strictures of the specific economic and political theory which he rather uncritically assumes, a balanced one. I shall attempt to show, however, that it does not go far enough. It fails to break free from its own ideological bias, even as it accuses AT of excessive programmatic zeal. The conversation concerning AT needs to make more of a break, to be more radical.

HARVEY BROOKS ON AT

Brooks characterizes his position as a synthesis of two opposing stances. On the one hand, he identifies the viewpoint that consciously rejects the practices of contemporary technology. E. F. Schumacher and Amory Lovins are identified as supporters of such a view.[15] It should be mentioned in passing that this identification is questionable in one important respect. Lovins is convinced that capitalism can be made to work using AT; in fact, that it will function more efficiently than current arrangements.[16] By contrast, Schumacher traces problems with current technology to wrong values built into market capitalism. Had he provided a coherent system of economic and technological principles alternative to market capitalism, instead of a series of semi-religious utterances, the AT movement could well have had something more permanent to show for its efforts.

Brooks also wishes to distance himself from those who dismissed AT out of hand as a romantic aberration when discussed in the context of the industrialized countries or who labeled it a "put down" when applied to the Less Developed Countries (LDCs). Schumacher has been criticized for advocating "intermediate technology" for the LDCs.[17] His recommendation was based on first-hand experience with Third World countries in which the transfer of capital-intensive, centralized technologies from the industrialized countries often led to massive breakdowns and long-term debt for the LDCs. It was often the elite of these countries that were the most critical of Schumacher's suggestions for technologies intermediate between the scratch plow and the assembly line factory.

Brooks's synthesis of these two positions may be called his "niche theory." He is persuaded that certain of the criticisms offered by the AT advocates possess real merit and have begun to affect normal practices. These reforms, however, are not large in number and certainly will not alter the majority of technological applications.

In effect what I am saying is that AT must be truly appropriate to the circumstances in which it is used, but that its appropriateness in comparison with more traditional technologies is likely to be quite limited geographically, socially, and ecologically. . . . [T]he proper place for AT is in 'niches' that are symbiotic with more conventional technology.[18]

Brooks's main points are as follows: *First*, proponents of AT adopt a *programmatic* approach to technological design. They explicitly promote the following design norms: small-scale systems, reduction of systemic interdependence, minimization of ecological impact, utilization of renewable resources, decentralized management, and labor-intensive but meaningful work.

Second, the adoption of such a program amounts to an *a priori* superimposition of ideology upon the normal way technological systems self-correct. Thus AT cannot play a major role because its ideological and programmatic definitions of "appropriate" interfere with the natural, and ideology-free, process of technical refinement.

Third, at the same time, AT *does* focus attention on technological practices in need of correction. While it is true that some imperfections in the conventional system, such as diseconomies of scale, are already self-correcting, others may benefit from the positive alternatives promoted by AT. In the "niches" AT will play a role where conventional practice is inefficiently correcting.

Brooks includes certain detailed observations to buttress his case: AT goals are, in some cases, inconsistent. For example, decentralization in transportation in the form of Ford's "tin lizzy" was only possible by increased centralization in the production process, that is, the assembly line. In other cases he sees AT goals as plainly wrong-headed. Calls for minimization of interdependence draw his most outspoken condemnation. "Interdependence is a fact of modern life."[19] While there is merit to this criticism especially when applied to those practitioners of the "back to nature" wing of the AT movement, Brooks misses the point of this complaint. Interdependence in the sense of community is quite different from interdependence in terms of a vast, impersonal network or bureaucracy. It is the latter that Lovins and Lovins, for example, have shown to be susceptible to catastrophic disfunction,[20] and which Winner describes as "reverse adaptation."[21]

Certain goals of AT, other than being inconsistent or plainly wrong-headed, are characterized by Brooks as superfluous. The correction process is already occurring by conventional methods. Criticisms of the scale of industrial technology and calls for decentralization fall into this category.[22] "It seems almost certain that . . . resistances to further increases in scale will lead to a greater diversification and dispersion of technology in the future."[23] One problem we are left with is understanding exactly how this self-correction is to be accomplished. The answer must not bear an "ideological" taint; that seems sure. He writes:

The cause of developing and diffusing AT would best be served by de-emphasizing its ideological aspects and working on the best possible adaptation of all technologies to their circumstances of use, irrespective of whether they fit the definition of "appropriate" or not.[24]

It is difficult to avoid the conclusion that Brooks genuinely believes that his own position is ideology-free. Yet, chiding the AT movement for choosing ideology in place of the "best possible adaptation of all technologies to their circumstances" clearly begs the question of what counts as appropriate or best.

Brooks does not fall into the economist's trap of adopting a sharp dichotomy between economic and social criteria but notes, rather, that social criteria are incorporated in the ground rules of the market itself. It is therefore proper to question the current mix of social criteria implicit in market economics just as the AT proponents do. What is not clear, he asserts, is that the programs promoted by AT uniquely embody these criteria, or that an enlightened evolution of conventional technologies might not achieve the same result.

We are offered little insight by Brooks as to how this might occur. We are told that "technological monocultures" appear to be reversing, that:

Realization of diminishing returns associated with large hierarchical organizations . . . has led . . . to the notion of sociotechnical design, the matching of technology to the human requirements of work and to more democratic and participatory work organizations.[25]

Where, one might ask, did these notions originate? Whereas the proponents of AT have attempted to deal consciously and explicitly with social criteria in the process of arriving at technological design norms, Brooks creates the impression that these steps toward improved technologies will probably just happen. In a statement that strains credibility and reflects a deep misunderstanding of the spirit of AT, Brooks states:

From the standpoint of social criteria technology may be regarded as a "black box." What is important is not what is inside the black box, but rather the "transparency" of the interface.[26]

Here lies the road to mystification! Instead of "technology with a human face" (Schumacher) or "convivial tools" (Illich), we are to concentrate on the most immediate, first-order effects of the latest technology. If current offerings fail to reflect desirable social criteria, we have but to wait for the "black box" to make corrections.

Brooks betrays an uncritical acceptance of the hedonistic conception of human nature implicit in the market model of humankind, a conception which, Veblen observed, renders us passive, substantially inert, a homogeneous globule of desire.[27] The reconfiguration of technological practice will require more than new artifacts. Instead, we may anticipate conversations which call into question the dominant atomistic, utilitarian conception of human nature. When one encounters an idea as obscure as Brooks's "black box" metaphor, it seems clear that we have drifted pretty far down stream from the original arguments and debatable positions about human "nature" that were offered at the beginning of the industrial era. That Brooks can regard his assessment of AT and conventional technology as free from ideology indicates how little remains of the vitality and conceptual depth of those earlier arguments.

CREATIVE REDESCRIPTION: PRAXICAL ALTERNATIVES TO AT

Advocates of AT have acted on the basis of intuitions that current technological practices are inappropriate responses to social and political realities of the end of the twentieth century. They have attempted by example to define new and "softer"[28] paths into the future. Unfortunately, their prescriptions have all too often exemplified the same hardware approach to reform as practiced by those they criticize. The cure, that is, for the mistakes of technology is more technology. At the same time, there have been reminders that technology, politics, and economics cannot be neatly compartmentalized. I recall the alarm that was expressed in the solar energy literature when it was realized that major energy producers were buying up the small solar hardware companies. Rather than providing the vehicle for the democratization of energy production and distribution – the dream of the solar advocates – the hardware was itself about to become one more instance of an oligopolistically controlled commodity.

One can find in the annual publications of the New Alchemy Institute an appreciation of the social role of technological devices. There are also manifest in these writings attempts to articulate alternative models of the human/tool relationship. They show a deep appreciation, for example, of the poor fit between analytical specialization demanded by the natural sciences, as presently constituted, and the systemic knowledge required to understand and to protect living ecosystems. Here they grope toward an alternative scientific methodology.[29]

These meritorious efforts do not, in my judgment, go far enough. Proponents of AT have been correct in their intuitions. Many of the current technological practices *are* out of true with our best intentions. Yet the intellectual frameworks used by the critics of AT provide little room for accommodation. Epithets such as "inefficient," "impractical," "utopian," "naive," "ideological," come all too easily to mind and do not conflict with a general impression that there is a lack of common sense displayed in the schemes of the AT promoters. How can these frameworks themselves, which are definitive of this "common sense," be made subjects of reflexive scrutiny? Philosophers may be able to act as facilitators in such a process, bringing to light what may otherwise remain inarticulate. Behind the intuitions about maladroit technology lie poorly developed ideas of what technology could be.

In a recent volume devoted to a non-foundational redescription of philosophy, Charles Taylor presents a strong case for philosophy being one and the same with the history of philosophy.[30] According to this Hegelian approach, certain problems can only be understood genetically.

[T]o understand ourselves today, we are pushed into the past for paradigm statements of our formative articulations. We are pushed back to the last full disclosure of what we have been about, or what our practice has been woven about.[31]

One such disclosure occurred at the beginning of the era of Western industrialized technology. For Hobbes and Locke, humans were pictured as atomistic, individual seekers of welfare, forming societies for the promotion of common interests and out of mutually experienced fears. For comparable prudential reasons contracts were entered into. It was against this background that the rationale of technological capitalism grew. The result was a creative redefinition of social life in commodity terms.

When we look beyond the modern practice of economics with its quantificational preoccupations and arcane theories to the origins of market economies, we are struck by how provisional and even arbitrary this new mode of thinking actually was. The emergence of an economics-based paradigm literally restructured the social world. The medieval connection between ethics and monetary transactions, reflected in the concept of "just price" was scuttled as were prescriptions against money lending (usury). Impersonal forces of supply and demand dictated availability irrespective of one's need or ability to pay. Feudal protections of serfs, however capricious and idiosyncratic they may have been, were simply abrogated as sheep replaced serf, and home, livelihood, and possessions were transformed into land, labor and capital.[32]

Where it occurred this monetization of social life was total. As Karl Polanyi notes:

A market economy must comprise all elements of industry, including labor, land, and money. . . . But labor and land are no other than the human beings themselves of which every society consists and the natural surroundings in which it exists. To include them in the market mechanism means to subordinate the substance of society itself to the laws of the market.[33]

This process Polanyi describes is characterized by the sociologist Robert Nisbet as the coming to the fore of a new social institution, the market economy, which began to compete with older institutions of family, clan, guild, religion, and nation.[34] No longer were matters of production or distribution of scarce resources worked out within the context of religious, social, and political custom and practices.[35]

When we come to appreciate the expulsive force of market economics in driving out ethical considerations of just price or obligation to serf as worker, we are better able to understand the ease with which contemporary economists and conventional designers dismiss AT proposals, however much of a "good thing" they may appear to be, simply by branding them as "uneconomic." Schumacher reminds us:

In the current vocabulary of condemnation there are few words as final and conclusive as

the word "uneconomic." . . . Call a thing immoral or ugly, soul-destroying or a degradation of man, a peril to the peace of the world or to the well-being of future generations; as
long as you have not shown it to be "uneconomic" you have not really questioned its right
to exist, grow, and prosper.[36]

The legitimation of exchange or manufacturing within a market
economy depends on strict and narrow considerations of economic
versus uneconomic practice. In another sense, however, Schumacher
goes on to show that this very restricted criterion is itself too broad.
"Once any goods, whatever their meta-economic character, have appeared on the market, they are all treated the same, as objects for
sale."[37] Yet, he points out, sounding a now familiar theme among proponents of AT, there are qualitative differences in goods, differences, for
example, between primary and secondary goods which ought to affect
practice but which are invisible to conventional accounting categories.
Secondary goods, manufactured items as well as services, require available primary resources, materials and energy. An expansion of production potential for secondary goods requires, as Schumacher notes, an
expansion of human ability to win primary products from the earth.
Furthermore, while the category of primary goods contains both renewable and non-renewable material and energy forms, our cost calculations involving price differentials alone completely obscure the fact that
non-renewables are a *stock* only while renewables comprise both *stock*
and *flow*. AT was explicit in arguing that the conflation of renewable
with non-renewable forms is pathological in the face of perceived
material and energy limits. Yet, how are we to treat this pathology?

Alternative economist Nicholas Georgescu-Roegen proposes that
something other than monetary cost be elevated to the position of
ultimate arbiter.[38] Orthodox market theory, he argues, fails to distinguish between stocks and flows of energy and materials. Instead of the
reversible and mechanistic model of economic supply/demand cycles
which characterizes the orthodox approach, he proposes an economic
model based on the second law of thermodynamics. This formulation
recognizes a direction to human activity. Every human intervention
expends more energy than is transformed into alternate forms of energy
or into ordered material configurations. By regarding each transaction
as an increase in terms of second-law entropy, a choice of renewable
flows over non-renewable stocks becomes preferable. Since every use of
a renewable resource makes use of the continuous flow of energy (solar)
while every use of a nonrenewable commodity does not, the former

represents a much smaller increase in entropy than does the latter. Georgescu-Roegen's program thus amounts to a new way of accounting in entropic terms that corrects the paradigmatic blindness of conventional economics to the inexorably increasing energy debt incurred in technological action.

It is just at this point that Charles Taylor's insistence on a genetic understanding of persistent, although possibly outmoded, cultural themes becomes particularly applicable. One cannot effectively escape the grip of a model, conventional market economics, for example, simply by envisaging an alternative as Georgescu-Roegen has done.

[F]reeing oneself from the model cannot be done just by showing an alternative. What we need to do is get over the presumption of the unique conceivability of the embedded picture. But to do this, we have to take a new stance towards our practices. Instead of just living in them and taking their implicit construal of things as the way things are, we have to understand how they have come to be, how they came to embed a certain view of things. In other words, in order to undo the forgetting, we have to articulate for ourselves how it happened, to become aware of the way a picture slid from the status of discovery to that of inarticulate assumption, a fact too obvious to mention.[39]

Taylor is not arguing that a return to the "last perspicuous formulation" of a theory, in our case commodity-based market economics, means that current practices are nothing but the embodiment of earlier explicit theories. He terms such a view "madly idealistic."[40] Nor are we to conclude that the original formulation which we are intent on comprehending is "somehow the true articulation of what underlies today's reality." Taylor specifically observes the wide disparity between the modern giant, bureaucratic, multinational corporation and the manufacturing firm envisaged by Smith in *Wealth of Nations*. Why, then, try to recapture the formative articulation?

We very often cannot raise a *new* issue really effectively until we have re-articulated our *actual* practices. But these frequently owe something to an outlook which was better, or fuller, or more perspicuously formulated in the past.[41]

If the prototypical outlook can be recaptured, we will have gained access to a social practice before "social change, drift, the pressure of other practices, unsuspected success, alterations in the scale of society,"[42] will have set in. We will have recaptured the optional nature of the controversy in question – in our case the question of whether a commodity basis of social life is defensible – when it was still an exciting case of creative redescription and before it became what is today "too obvious for words."[43]

Taylor's position helps explain why the proposals of AT were so easily dismissed or simply ignored, why they had no bite. When a striking new way of looking at things gains acceptance and becomes an established social practice, a process of "historical forgetting" occurs. The model becomes embedded in the ways we think, act, and deal with the world and thus sinks to the level of unquestionable background assumption.

Challenging an outlook of this kind means undoing this process of forgetting. Just presenting an alternative will not help, as long as we are captured within the received "common sense." For if this holds us in virtue of being embedded in our practices, then we have to articulate what they embody if we are to neutralize its effects. Otherwise we remain as it were captured in the force field of a common sense . . . [which] distorts the alternatives, makes them look bizarre or inconceivable.[44]

Such, it seems to me, describes the AT movements of the seventies. They were not reflexive. There was a flavor of the bizarre about them. Schumacher's provocative challenges to usage of the term "economic" were never placed within a coherent framework but remained instead an uneasy mixture of Buddhism and Catholicism. His pronouncements too often took on an oracular flavor, abetted by a throng of followers who hung on every word spoken by their prophet for the coming age. Another noteworthy advocate of AT, John Shuttleworth, founder of the *Mother Earth News*[45] in Hendersonville, North Carolina, promoted a survivalist position which emphasized self-sufficiency reminiscent of nineteenth century anarchist movements. Shuttleworth was disdainful of scholarly analyses and so was largely ignorant of the predecessors of his own life style and virtually oblivious to the philosophical positions on which his "common sense" opinions were based.

Amory Lovins, who coined the phrase "the soft path,"[46] attempted, with copiously documented evidence, to demonstrate that AT was in fact truer to the real spirit of capitalism than were the centralized, large-scale enterprises of our times. Such a point of view ties in nicely with Brooks's "niche theory" of evolutionary development of existing technological practices. Despite its attention to quantitative detail, however, Lovins's view is critically superficial, an analysis that is largely unaware of its own "historical forgetting."

CONCLUSION

If we are interested in the reform of contemporary technologies we cannot simply promote different tools. Because artifacts embody forms of life, as Winner so aptly puts it,[47] it is dimensions of culture embodied in such forms that must be appropriated and made relevant to contemporary concerns. We have suggested that this task can require an historical re-creation of controversies that were once deadly serious but are now kept alive as mere academic talking points. Is it possible, for example, to discuss the role of technology in culture in non-commodity-based terms? Can the context of discussion be recaptured in which Smith's model of a market economy is seen as a brilliant, if controversial, solution, and not simply something so obvious as to be beyond doubt? If, as Winner also observes, we once before in our history, at the onset of the modern industrial era, passed up the opportunity to engage in discussions of the forms that industrial technology should take,[48] will we be any more successful this time around?

This discussion has adopted a critical and negative position regarding such a possibility if the approach consists primarily in the mere fashioning of alternative tools. Such was the largely ineffectual approach of AT. A different strategy aims to promote a genetic understanding of today's "common sense" about the role of technology in society. Albert Borgmann, in my opinion, exemplifies such an approach with his development of the "device paradigm."[49] By returning to the beginnings of market economies, he is able to pinpoint the diremption of social and political life into machinery, on one hand, and commodities, on the other. The device thus becomes the means whereby social life is realized and in which "liberal democracy is enacted as technology."[50] Here, by going to the historical origins of the issue, Borgmann offers insight regarding the technological milieu in which we are embedded. He has thus realized, in one instance, a creative redescription such as Charles Taylor has called for.

The praxical, non-foundationalist analysis of technology which I have adopted herein rejects the idea that the philosopher, or any other member of a cultural community can find an Archimedean point outside that culture from which to legislate *the* essential meaning or *the* appropriate form of technology. Because its concrete instances comprise "forms of life," its use both shapes social practices and changes as those practices evolve. We can better understand the present by means of a deeper insight into the past. We cannot, however, escape cultural categorizations altogether. Philosophical analysis must therefore be practiced within the hermeneutic circle. A certain circularity of reason is thus unavoidable.

When we attempt to arrive at a concept of appropriate technology that is transcultural or cognitively independent of particular embodiments, we betray the project of philosophy as conversation. To attempt to define the limits of what is appropriate technological practice in essentialist terms amounts to a move to silence criticism. Instead of repeating the style of philosophical argumentation that seeks to construct arguments which are so devastatingly compelling that one must *necessarily* accept them – philosophy by force instead of by rhetoric, that is – a praxical approach can, at best, recapture, systematize, or challenge transient conceptions of what it is good for us to do with our hands. We cannot, thus, provide *a priori* the formal ground rules for adjudicating controversies between rival theories of appropriate technology. Our forte remains our way with words and this facility is what we can contribute to the ongoing, self-defining conversations concerning the human project.

Georgia Institute of Technology

NOTES

[1] The phrase "appropriate technology" is often restricted in its application to considerations of industrial development in "Third World" countries. The increasingly maligned phrase, "intermediate technology," coined by E. F. Schumacher (see 'Social and Economic Problems Calling for the Development of Intermediate Technology', in *Small Is Beautiful: Economics as if People Mattered*, New York: Harper & Row, 1973, pp. 161–179) is frequently used synonymously. Initially, at least, Schumacher's point was that practices that had evolved in the industrialized countries were ill-suited to less-developed countries which often lacked basic amenities, social institutions, and infrastructures such as roads, schools, telephone, internal sources of capital, repair facilities and technical know-how. A technology somewhere between the scratch plow and the sophisticated diesel tractor, he argued, was more appropriate. Schumacher's incisive critique was adopted by others and broadened in scope to include the ensemble of technological practices, found both in the LDCs and in the industrialized countries. It is in this broader context that I apply the term here.

[2] The ideas discussed in this paper were first raised in 'Technology Assessment without Mirrors: Deconstructing Foundationalist Philosophies', presented at Inter-University Centre of Postgraduate Studies in a course devoted to 'Social Interpretation of Technic', Dubrovnik, Yugoslavia, 2–13 April 1984. The current version is a revision of a paper prepared for the third international meeting of the Society for Philosophy and Technology, Technical University at Twente, The Netherlands, 12–16 August 1985.

[3] I have been particularly influenced by Richard Rorty's line of argument in *Philosophy and the Mirror of Nature* (Princeton: Princeton University Press, 1979) and *Consequences of Pragmatism: Essays 1972–1980* (Minneapolis: University of Minnesota Press, 1982).

See also R. Rorty, J. Schneewind, Q. Skinner, eds., *Philosophy in History: Essays on the Historiography of Philosophy* (Cambridge: Cambridge University Press, 1984), esp. articles by C. Taylor, pp. 17–30; R. Rorty, pp. 49–75; and A. MacIntyre, pp. 31–48.
[4] Langdon Winner, *The Whale and the Reactor* (Chicago: University of Chicago Press, 1986), p. 80.
[5] Proponents are fond of making lists of the design goals that characterize AT. One of the most comprehensive and often quoted lists was prepared by Robin Clark of Biotec Research and Development (BRAD). Clark lists thirty-six dimensions of contrast between the "hard" technology approach and the "soft" technology programs. In general, the term "soft" refers to soft or gentle effects of the technology in question on the social and physical environment in which it is introduced. One of the most accessible sources of Clark's list is given in Dickson's excellent book, *The Politics of Alternative Technology* (New York: Universe Books, 1975), pp. 103–104.
[6] Winner, *op. cit.*, note 4, p. 82.
[7] *Ibid.*, p. 80.
[8] A Benthamite move from descriptive to normative would have been accomplished, the "naturalistic fallacy" notwithstanding. Practical difficulties with such a program are admittedly formidable. How to optimize multiple preference variables is, it is true, a difficult technical question. Risk/cost/benefit is one approach with promise. Its refinement poses major problems particularly involving equity preference functions. Presumably satisfactory solutions await systematic grounding in ethical theory, such theories themselves being essentialist in character.
[9] W. V. O. Quine, 'Two Dogmas of Empiricism', in his *From a Logical Point of View* (Cambridge, MA: Harvard University Press, 1953), pp. 20–46.
[10] Thomas Kuhn, *The Structure of Scientific Revolutions* (2nd ed.; Chicago: University of Chicago Press, 1970).
[11] Responding to criticisms that the term "paradigm" is used in a variety of ways in *The Structure of Scientific Revolutions*, Kuhn asserts in the 'Postscript' to the second edition that all usages of the term may be grouped under the categories of "disciplinary matrix" and "exemplars" of particular theories; see pp. 174–175.
[12] Joseph Margolis, 'Information, Artificial Intellience, and the Praxical', paper presented at the second international conference of the Society for Philosophy and Technology on information technology and computers, Tarrytown, N.Y., September 1983; see C. Mitcham and A. Huning, eds., *Philosophy and Technology II: Information Technology and Computers in Theory and Practice* (Dordrecht: Reidel, 1985), p. 171.
[13] Langdon Winner, 'The Political Philosophy of Alternative Technology: Historical Roots and Present Prospects', in D. Lovekin and D. Verene, eds., *Essays in Humanity and Technology* (Dixon, IL: Sauk Valley College, 1977), p. 131.
[14] Harvey Brooks, 'A Critique of the Concept of Appropriate Technology', in F. Long and A. Oleson, eds., *Appropriate Technology and Social Values: A Critical Appraisal* (Cambridge, MA: Ballinger, 1980), pp. 53–78.
[15] Amory Lovins, *Soft Energy Paths: Toward a Durable Peace* (Cambridge, MA: Ballinger, 1977). This is an expansion of his much discussed paper, 'Energy Strategy: The Road Not Taken', *Foreign Affairs* **55** (October 1976): 65–96. See also Schumacher, *op. cit.*, note 1.
[16] Lovins, 'A Neo-Capitalist Manifesto: Free Enterprise Can Finance Our Energy Future', *Politicks & Other Human Interests*, **1** (April 11, 1978): 15–18.

[17] Schumacher, 'Social and Economic Problems Calling for the Development of Intermediate Technology', *op. cit.*, note 1, pp. 161–179.

[18] Brooks, *op. cit.*, note 14, pp. 73, 75.

[19] *Ibid.*, p. 68.

[20] Amory Lovins and L. Hunter Lovins, *Brittle Power: Energy Strategy for National Security* (Andover, MA: Brick House Publishing Company, 1982).

[21] Langdon Winner, *Autonomous Technology: Technics-out-of-Control as a Theme in Political Thought* (Cambridge, MA: MIT Press, 1977), pp. 226–236.

[22] See my 'Scale in Technology: A Critique of Design Assumptions', in P. Durbin ed., *Research in Philosophy & Technology*, vol. 8 (Greenwich, CT: JAI Press, 1985), pp. 67–76.

[23] Brooks, *op. cit.*, note 14, p. 59.

[24] *Ibid.*, p. 77.

[25] *Ibid.*, p. 61.

[26] *Ibid.*, p. 55.

[27] Thorstein Veblen, 'Why Is Economics Not an Evolutionary Science?' in *The Place of Science in Modern Civilization and Other Essays* (New York: Huebsch, 1919), p. 73.

[28] Lovins observes, "For lack of a more satisfactory term, I shall call them 'soft' technologies: a textural description, intended to mean not vague, mushy, speculative, or ephemeral, but rather flexible, resilient, sustainable, and benign." Social structures utilizing soft technologies are characterized by Lovins according to five criteria: (1) reliance upon renewable energy flows; (2) diversity of approaches; (3) utilizing flexible and relatively low-technology systems; (4) matched in scale and in geographic distribution to end-use needs, taking advantage of natural energy flows; and (5) matched in energy quality to end-use needs. See *Soft Energy Paths*, pp. 38–39.

[29] "The New Alchemy Institute was founded in 1969 by two marine biologists, John H. Todd and William O. McLarney. The choice of name implied not a rejection of modern science but a harking back to a time when science, art, and philosophy did not have to be practiced as separate, mutually exclusive realms of knowledge. Todd had a broader training than most scientists – a B.Sc. in agriculture, an M.S. in parasitology and tropical medicine, and a Ph.D. in comparative psychology and ethology – yet he and his associates found, he says, that 'with all our scientific training, we could not make any little piece of the world work'" – *Science*, **187** (1975): 727. The New Alchemy Institute now has centers in Woods Hole, Mass., and Prince Edward Island, Canada. They publish a beautifully illustrated annual volume, *The Journal of the New Alchemists*, available from P. O. Box 47, Woods Hole, MA 02543.

[30] Charles Taylor, 'Philosophy and its History', in *Philosophy in History: Essays on the Historiography of Philosophy* (see note 3, above), pp. 17–30.

[31] *Ibid.*, p. 26.

[32] Robert Heilbroner, *The Making of Economic Society* (5th ed.; Englewood Cliffs, NJ: Prentice-Hall, 1975), esp. pp. 23–46.

[33] Karl Polanyi, *The Great Transformation* (New York: Rinehart, 1944), p. 71.

[34] Robert Nisbet, 'The Impact of Technology on Ethical Decision-Making', in his *Tradition and Revolt* (New York: Random House, 1968), p. 186.

[35] *Ibid.*

[36] Schumacher, 'The Role of Economics', in *Small Is Beautiful*, pp. 39, 40.

[37] *Ibid.*, p. 47.

[38] Nicholas Georgescu-Roegen, *The Entropy Law and the Economic Process* (Cambridge, MA: Harvard University Press, 1971).

[39] Taylor, *op. cit.*, note 30, p. 21.

[40] *Ibid.*, p. 25.

[41] *Ibid.*, p. 24. Emphasis in original.

[42] *Ibid.*, p. 25.

[43] *Ibid.*, p. 20.

[44] *Ibid.*, p. 24.

[45] Shuttleworth came to Georgia Tech in 1974 to enlist architecture design students in the design of a self-sufficient ecological village. The students produced a number of novel layouts using solar and bio-mass energy generation. To my knowledge, the village never became a reality although one can visit the headquarters of the *Mother Earth News* and witness a number of innovative examples of alternative technology, such as a truck that runs on animal waste.

[46] Lovins, *Soft Energy Path*.

[47] Winner, 'Technologies as Forms of Life', in *The Whale and the Reactor*, note 4, pp. 3–18.

[48] Winner, 'The Political Philosophy of Alternative Technology', note 13, p. 135.

[49] Albert Borgmann, *Technology and the Character of Contemporary Life* (Chicago: University of Chicago Press, 1984), pp. 40–48.

[50] *Ibid.*, p. 90.

THOMAS SIMON

SYMPOSIUM ON APPROPRIATE TECHNOLOGY
II. Appropriate Technology and Inappropriate Politics

INTRODUCTION

Over the past two decades "appropriate technology" has been used as code for new ways of thinking about the social implications of technological choice. Appropriate technology is seen variously: as a means of ushering in a New Age, as an alternative to high technology, as a social movement, and, by some, as utopian delusion. The debate over appropriate technology is raising some of the most difficult questions facing a philosophy of technology, including:

The relationship between technology and development, between ideology and industrialization, and more fundamentally, between man and machine (Rybcynski, 1980, p. v).

Yet, the most critical question is seldom explicitly addressed in this debate: what is and what should be the political philosophy underlying the appropriate technology movement?

Before providing my own answer to that question we need to get clear about just what appropriate technology is. Some critics charge that appropriate technology is a grab bag of vague ideas, which resist any attempts to provide definitional coherence. According to this view,

[Appropriate] technology is not a coherent philosophy. It is a collection of a large number of ideas and concepts, many of them quite incoherent, almost as diverse as the name of the outlook: intermediate technology, humane technology, a new alchemy, peoples' technology, radical hardware, biotechnics, etc. Each of these names emphasize a different aspect of the new technology: workers' control, demystification of expertise, reform of work rules, low specialization, development under the condition of low capital, local or regional self-sufficiency, balanced economic development, resource conservation, low energy use, reduced technological risks, and so on (Sardar and Rosser-Owen, 1977, p. 564).

Contrary to this assessment, I will argue that methodological coherence can be given to the appropriate technology (AT) proposals. To do this, I will first look at what proponents of AT indicate as their agenda, and then I will undertake a rational reconstruction of the aims and assumptions of this agenda. In other words, prevailing AT ideas and practices can be used to define a methodologically adequate concept of AT.

Armed with a clear definition of AT we should be in a position to

107

Paul T. Durbin (ed.), Technology and Contemporary Life, 107–128.
© *1988 by D. Reidel Publishing Company.*

differentiate appropriate technology from its close kin – ecotechnology, liberatory, indigenous, intermediate, and radical technologies. Each of these might be seen to emphasize a different definitional feature than AT emphasizes. This approach, however, will be shown to be misleading.

In the final section of the paper I will demonstrate that the problem with AT (and its kindred conceptions) is not methodological adequacy but rather the problem is the political foundation of the particular philosophy of technology. Appropriate technologists debate high technologists in terms of the features of their differing technologies. Likewise some ecotechnologists, liberatory technologists, etc., debate appropriate technologists in terms of the features of their differing technologies. This, however, is the wrong debate. The debate should not be over the features of the technologies. The debate is covertly and should be overtly a political debate. Failure to recognize the political nature of the debate leads critics to see a hodgepodge among these differing conceptions of technology. Recognizing the political nature of these different conceptions results in AT being judged politically inadequate. Before making the case for its political inadequacy let us first demonstrate the methodological adequacy of AT.

SECTION I

Defining Appropriate Technology

When you read the AT literature and come across definitions of appropriate technology such as "technology with a human face" (Schumacher, 1973) and "whatever is appropriate" (Ndongho and Angang, 1981), you can immediately see the need for some definitional clarity. Those AT proponents with some methodological sophistication use three definitional tactics: providing ostensive definitions, listing evidentiary features, and specifying necessary and sufficient conditions. While I will argue that the last tactic is the most convincing, the others are worth describing since the discussion sheds further light on the AT movement.

A. *Ostensive definitions*

Some appropriate technologists readily concede the difficulties in providing a clear definition of AT: "Although we do not know how to define it, we do know an appropriate technology when we see one" (Jequier, 1976). There is something to this attitude. After all, whatever the definitional complexities, the differences between a bulldozer and an ox-plough or between a nuclear power plant and kerosene lanterns are readily apparent.

Nevertheless, there are cases where we cannot simply see the difference between an appropriate technology and other forms of technology. What, for example, is the most appropriate technology for maize-grinding in Kenya – mortar and pestle, hand operated mills, hammer mills, or roller mills (Stewart, 1978, chapter 9)? Making this determination by simply looking at the candidate technologies is doomed from the outset.

The ostensive-definitional strategy reflects an attitudinal bias which needs to be borne in mind when assessing AT. The attitude manifests itself in the impatient "Let's get on with it" directive typically issued by technologists of all varieties. Generally, the technologist wants to give priority to doing over knowing. Problems cannot await the plodding intellectualization of philosophers and theoreticians. It is important to note that there are many proponents of appropriate technologies who subscribe to this attitude. This should underscore the point that appropriate technologists are first and foremost technologists, a point which is developed further in this paper.

While I have some sympathy with the technologists' attitude, this attitude rules out a certain fundamental line of inquiry. If I say "Let's get on with it" to practically any kind of project, I am thereby cutting off the questioning of the project itself, its underlying assumptions and implications. For appropriate technologists this can involve a refusal to examine the broader social/political context of technology introduction.

The tactic of providing ostensive definitions avoids the task of developing a political philosophy of technology, implying that techniques, at least "appropriate" ones, are politically benign and, perhaps, inherently good. This political avoidance behavior is also evident, although less blatantly so, in the other definitional approaches. Before showing that, I want briefly to describe a "Let's get on with it" attitude among philosophers, comparable to the one found among technologists. Philosophers have little patience with extended treatments of ostensive definitions since they are so quickly dismissed in the philosophical literature. However, what is obvious in one discipline is not so obvious in another. Many practically oriented technologists take ostensive definitions seriously, and philosophers should at least be willing to give the technologists the benefit of the doubt.

B. *Evidentiary features*

Other commentators offer as a definitional strategy a shopping list of features which, supposedly, can be expected to be found in the appropriate technology store. No one feature or set of features is regarded as necessary or sufficient for appropriate technology, but the addition of each feature is argued to increase our inductive confidence that a candidate technology is appropriate. A solar satellite would presumably have relatively few of these features, while active solar heating and, certainly, passive solar design can be comfortably situated in the AT camp. What are these discriminating features?

No single list is found in the AT literature. Huelphinil proposes fifteen features (de Wilde, 1977, pp. 162–163); Robin Clark proposes thirty-five, the longest list. Both lists seek to distinguish an appropriate technology society from a hard technology one. An AT-organized society emphasizes: functionality for all time, communal units, local bartering, and integration of young and old. In contrast, hard technology breeds: functionality for a limited time only, nuclear families, worldwide trade, and alienation of young from old. Diwan and Livingston (1978, p. 93), trying to order the chaos, group the features from various lists in the following way:

In terms of material aspects of AT production, "Appropriateness" connotes the use of renewable sources of energy and recyclable materials, minimum destructive impact on the environment, and maximum utilization of local resources. In terms of the modes of production, AT fabrication should take place close to the resource base, using processes which are labor intensive (capital saving), small scale, amenable to user participation and/or worker management, and located close to points of consumption. In terms of application, AT connotes assimilation with local environmental and cultural conditions. It does not overwhelm the community, but is comprehensible, accessible, and easy to maintain.

Despite some progress achieved by this grouping, the list approach in general confronts a number of methodological problems. One is "how incompatible [features] are to be combined" (van Brakel, 1980, p. 392). Let us take two features from Diwan and Livingston's list, *viz.*, minimum destructive impact on the environment, and small scale. Rybczynski (1981, p. 20) proposes "the flush toilet, the throwaway container, the aerosol can, . . . and deforestation" as counterexamples to the claim made by AT proponents, like Schumacher, that small scale devices are environmentally benign compared to large scale ones. Small scale devices can have a considerable negative environmental impact. (Later we will see how a better definitional strategy can meet this objection.)

Moreover, some items that we might find intuitively unacceptable as appropriate technologies qualify under the list approach. For example, do we find a description of an appropriate technology in the following?

It will be large enough for the family but small enough for the individual to run and care for. It will be constructed of the best materials, by the best men to be hired, after the simplest designs that engineering can devise. But it will be so low in price that no man will be unable to own one – and enjoy with his family the blessing of hours of pleasure in God's great open spaces (quoted in Rybczynski, 1980, p. 19).

Appropriate technology features abound in this description: small scale, simplicity, economical, amenable to mastery and maintenance by local people, etc. Yet, this was Henry Ford's description of the universal car, the Model T, hardly a fitting exemplar for AT enthusiasts. Similar descriptions can be found among nuclear power proponents, who are typically arch-enemies of AT advocates. As an example take the 1946 claim of Weinberg, Wigner, and Young, that "nuclear power plants [would] make feasible a greater decentralization of industry" (as quoted in Khalilzad and Bernard, 1980, p. 17).

Again, as in the case of ostensive-definitional tactics, something more than methodology is amiss here. Winner (1980, p. 40) correctly derides the paradise-for-Christmas *versus* the sordid-perils-of-modernity lists as shallow social criticism. Oddly enough, all the good things in life are found under the appropriate technology tree while all the evils group themselves under the "high technology" label. Just as we found beneath the "Let's get on with it" technologist's impatience with theory a refusal to deal with social and political issues, so we find here a social-political criticism disguised as a debate over types of technologies. When proponents of appropriate technology present these lists, they presuppose both a particular critique of current society and a proposal for a better society. Yet, the critique and proposal are never made explicit. Instead, we are to accept or reject them in terms of technological choices.

One final methodological problem besetting the evidentiary feature strategy is the inability to determine which features are central. For example, is using local resources, which results in destroying a rain forest, more important than environmental preservation? In the next section we will examine a strategy that helps overcome some of the methodological problems apparent in the other two strategies. Yet, this third strategy, while making some progress, also fails to confront the problem of providing a political foundation.

C. *Conditions*

The following is a sample of typical definitions of appropriate technology in the literature:

– Technology which is decentralized, small in scale, labor intensive, amenable to mastery and maintenance by local people, and harmonious with local cultural and environmental conditions (Shaikh, 1981, p. 26).

– We emphasize four criteria: smallness; simplicity; capital cheapness; and non-violence (McRobie, 1981, p. 36).

– Besides its [AT's] ability to offer to every member of society the fullness of the *summum bonum*, it offers to everyone also the dignity of work needed to attain it (Kohn, 1980, p. 190).

– It is conducive to decentralization, compatible with the laws of ecology, gentle in its use of scarce resources, and designed to serve the human person instead of making him the servant of machines (Schumacher, 1973, p. 145).

– AT as content has come to be applied to a special class of technology as systems: those incorporating energy-efficiency, labor intensive, small scale, decentralized technologies (DeForest, 1980, p. 13).

From this widely variegated list, it is easy to sympathize with the charge that AT is an incoherent hodgepodge. Nevertheless, an adequate definition, acceptable to most AT proponents, can be salvaged from these proposals. If the above proposals are examined not as an unmanageable list of features but as alternative sets of conditions, it is possible, through selecting the best set of conditions, to devise an adequate definition of AT. However, as I will try to show, this definition, despite its methodological adequacy, falls far short of providing a much-needed political critique. Moreover, when the political foundations are fleshed out, they are found wanting.

My strategy in this section is to focus on DeForest's proposed set of conditions. Why I have chosen DeForest's set will become more evident when a final assessment of the definition is given in the next section. At this stage, suffice it to say that DeForest provides a manageable list of conditions, which overlap to some extent or another with almost every other definitional proposal of AT that I have seen. Nevertheless, De-

Forest's list needs fine-tuning. Each condition is therefore examined to determine how the set might contribute to a manageable and defensible framework for AT. The final step is to develop my critique of AT in terms of its being either politically impotent or politically inadequate.

1. *Energy Efficiency*: This condition is central to two of AT's kindred spirits, soft technologies and biotechnics. The key feature of these technologies is (unlike their counterparts, hard technologies) that "they are matched in *energy quality* to end-use needs" (Lovins, 1977, p. 39). In accord with this analysis, nuclear power is a highly inefficient means of heating homes since the end-use does not require such a high-grade form of energy. Similarly, home water heating is better done with natural gas than electricity since natural gas heats the water directly whereas a conversion process is required to produce electricity. Mismatching end-use typifies hard energy technologies.

2. *Environmental Soundness*: However valuable an analysis in terms of energy efficiency might be, it is not broad enough to capture the concerns of many AT proponents. Energy efficiency is important to appropriate technologists but not as important as environmental soundness. For example, the insulating and ventilating properties of Third World homes is not an issue for AT proponents. In contrast, there is a great deal of discussion within the AT literature about the conceivable environmental impact of introducing tractors into a region (Date, 1984). Accordingly, a technology is evaluated in terms of its impact on the immediate environment. Therefore, there are good grounds for rejecting energy efficiency as too narrow a condition for defining AT. Furthermore, where energy efficiency is a consideration, it can be subsumed under environmental soundness.

3. *Labor Intensity*: Almost every proponent of AT includes some form of labor – as opposed to capital – intensity as a minimum requirement (i.e., as a necessary condition) for AT. For example:

Instead of concentrating on labor-saving devices, which has been the whole trend of modern technology, can you turn your attention to capital-saving devices, because it is capital that is lacking in developing countries not labor (McRobie, 1977, p. 4).

Nuclear power and large farm tractors are capital intensive technologies, resulting in few workers employed. AT proponents oppose these options when other technologies can be deployed to carry out work which expands employment.

Because more people are employed, the benefits of growth will be spread more widely, and this wider distribution of income will contribute greatly to sparking demand for marketable goods in other industries (Jackson, 1984, p. 80).

(The founder of the AT movement, E. F. Schumacher (1973), makes the case for a Buddhist economics which is not preoccupied with marketable production. A thorough reading of the AT literature, however, reveals that Schumacher is an exception on this point.)

4. *Small Scale*: Schumacher outlined, in *Small Is Beautiful* (1973), the parameters of the AT movement, which McRobie sought to implement in *Small Is Possible* (1981). Since the publication of Schumacher's book, homage is repeatedly paid to the virtues of smallness; bigness takes on an almost immoral character. But recognizing the importance of smallness for AT is one thing; getting a clear sense of what smallness means is quite another. Schumacher regards smallness as a relative concept, necessary to offset the currently reigning "idolatry of giantism." Beyond that consensus smallness is used ambiguously throughout the AT literature to apply to production (Dickson, 1975, p. 106), equipment (McRobie, 1981, p. 36), or markets (Stewart, 1978, pp. 102–103). AT proponents see these three senses of smallness in terms of factors enabling people to attain control and as meeting the demands of the market.

5. *Decentralization*: Again, it is difficult to find a clearcut meaning of this term. Does decentralization mean that production units are widely dispersed spatially rather than concentrated in special regions? Does it mean putting factories in the rice paddies (Lappé and Collins, 1978, p. 470)? Or does it mean that decisions about work and production are to be made at the lowest levels of production? AT proponents use "decentralization" in two senses, *viz*., as a decision about where the technology should be placed, and as a decision about worker control.

6. *My Definition*: Summarizing the analysis so far, a modified version of DeForest's definition can be proposed. Appropriate technology will be taken to incorporate the following characteristics:

(1)   environmental soundness;

(2)   labor intensity;
(3)   small scale; and
(4)   decentralization.

These are conjunctively sufficient conditions for defining AT with no one condition sufficient in itself. Also, these conditions are disjunctively necessary for AT. Thus, this is a coherent set of conditions which capture (for the most part) what AT proponents mean when they talk about AT.

D. *Adequacy of the definition*

The proposed definition more or less captures what most proponents are trying to express about the features and qualities of AT. As a further defense of the modified DeForest proposal, I would argue that any proposed alternative condition would be either vaguer than any of the accepted conditions, subsumable under one of those conditions, or objectionable in its own right. For example, McRobie's claim that AT should be nonviolent is vague. Labor intensive technologies are easy to recognize: nonviolent ones are difficult to discern.

Many other proposed conditions can be subsumed under one or more of the accepted ones. For example, operating according to the accepted conditions will involve the use of renewable sources of energy and maximum use of local resources. Furthermore, adhering to these conditions generally yields a flexible, low-cost technology with a rural emphasis. In fact, many AT advocates use small scale as a condition because it is thought to result in a rural emphasis.

Finally, other candidate conditions can be readily dismissed. "Simplicity of operation" is just too problematic. The division of labor certainly simplifies operations. Are assembly lines, then, models of appropriate technology? I hardly think so. In summary, the accepted conditions include most of the features AT proponents want and exclude the undesirables.

More importantly, these conditions enable AT theorists to rebut previously telling charges. A number of critics argue against AT by taking each condition in isolation and then showing how that condition is not acceptable. This tactic has been used, for example, against the small-is-beautiful thesis. The smallness of a production process does not prevent environmental and human destruction on a massive scale. However, this violates the environmental soundness condition. Similarly, the criticism that a small slave system is more environmentally sound than a large factory farm is refuted by noting that a slave system violates the decentralization condition, giving workers some control over the process.

Also, some critics have sought to pit the conditions against one another, showing their mutual inconsistency (Winner, 1980). As noted above, Rybczynski (1981, p. 21) proposes the aerosol can and deforestation as counterexamples to Schumacher's claim that small scale devices are always more environmentally sound than large scale ones. Brooks (1980) follows a similar line, arguing the incompatibility of environmental soundness and decentralization. Response to these criticisms is straightforward. While the conditions are related (small scale is generally, but not always, more environmentally sound than large scale production), they are interdependent conditions that must be explained together when evaluating a particular technique. To take one of the alleged counterexamples, the aerosol can does not qualify as AT because it violates the condition of environmental soundness, since it is partially responsible for ozone depletion.

Thus far, I have demonstrated that prevailing AT ideas and practices can be used to define a methodologically adequate concept of AT. At this stage, it should be at least clear what people are talking about when they use the term "appropriate technology." From a definitional vantage point, AT can avoid the charge of representing an incoherent hodgepodge by appeal to a relatively clear set of conditions. Accepting that a defensible definition of AT can be rendered, we are now in a position to examine more fundamental criticisms of AT. Difficulties with AT are found, not in specifying what types of technology qualify as AT, but in constructing adequate philosophical, particularly political, foundations of AT.

SECTION II

The Political Foundations of AT and Related Technologies

The real problem with AT is not methodological. The problem is political. On the one hand, Winner (1986) is correct in claiming that technological choices are, in fact, political.

Choices about supposedly neutral technologies – if "choices" they ever merit being called – are actually choices about the kind of society in which we shall live (Winner, 1986, p. 315).

On the other hand, Winner is incorrect in thinking that AT offers an alternative political philosophy. AT proponents generally explicitly refuse to address political questions. To the extent that they do, AT political philosophy is not altogether that different from the political philosophy espoused by high technology advocates.

There are exceptions to this political failure within the AT movement. However, within AT that debate often takes on the character of a technological, not a political, debate. Ecotechnology, liberatory, indigenous, and intermediary technologies are thought to emphasize different features of the technologies. So, Sardar and Rosser-Owen (1977), quoted at the outset of this paper, are partially on target in criticizing these formulations for simply emphasizing different aspects of the new technology. Some of the proponents of these alternative formulations are guilty of that. But what they are really trying to do (some self-consciously) is construct an alternative philosophy of technology, emphasizing particularly the political aspects.

In what follows, I will try to articulate that philosophy in the course of showing that the debate between AT and these related technologies has been miscast as a debate over features of technology. Although I will examine each condition separately, my purpose is not to isolate them unwittingly but rather to reveal the political complex underlying each. I will be contrasting aspects of political philosophies, not features of technologies.

A. *Environmental soundness vs. ecotechnology*

It is not simply a quirk in linguistic habits when AT proponents speak of environmental rather than ecological impact. The choice in phrasing reflects a specific political philosophy. Yet, even AT's commitment to environmental concerns is questionable. Environmental concerns are given lip service, but seldom is there an extended description of actual environmental impact in AT case studies. But let us assume environmental awareness in AT practices.

In coining the term "ecotechnology," Bookchin (1980) makes an important distinction between the environmental and the ecological, not as features of the technology but rather as philosophies. Environmentally, a technology is regarded as a relatively isolated device whose impact on the environment can be more or less precisely calculated. Ecologically, a technology is regarded as "functionally integrated with human communities as part of a shared biosphere of people and non-human life forms" (Bookchin, 1980, p. 109). Bookchin reveals the important philosophical difference between these when he characterizes the concern of AT proponents over the impact of technology as

... within the context of environmentalism which tends to reflect instrumentalist sensibility in which nature is viewed merely as a passive habitat, an agglomeration of external objects and forces, that must be made more serviceable for human use irrespective of what these uses may be (Bookchin, 1980, p. 58).

When environmental issues are discussed in the AT literature, they are couched in this instrumental framework and not in a broader ecological context, whose conscious goal is promoting the integrity of the biosphere.

Environmental soundness and ecological integrity are not two comparable features of technologies. Underlying each are different political philosophies. AT proponents, for the most part, do not challenge the existing political power structure except for advocating some degree of popular control over the technology. According to Bookchin (1982, p. 243):

To speak of "Appropriate Technology" . . . without *radically* challenging the political "technologies," the media "tools," and the bureaucratic "complexities" that have turned these concepts into elitist "art forms" is to completely betray their revolutionary promise as a challenge to the existing social structure.

In other words, without an explicit alternative political philosophy AT offers little alternative beyond accommodation.

B. *Labor intensiveness vs. liberatory technology*

While putting people to work is a noble undertaking, the labor intensity condition becomes problematic when AT is compared to liberatory technology, also developed by Bookchin (1971). A primary liberating function of technology is that it can free people from toil and drudgery. Bookchin bemoans the ineffectiveness of earlier revolutionary movements in their inability to alter conditions and structures of scarcity. Failures of the past might be avoided, he argues, as we enter a post-scarcity phase of history where technology is available to relieve people of having to slave for subsistence.

Some of Bookchin's liberatory technology examples are highly debatable, for example, controlled thermonuclear reactions for mining, and genetic improvements of food plants. It is unclear how truly liberatory these are, and I have other reservations about his analysis. Notwithstanding challenges that can be made to his liberatory technology philosophy, it seems clear that AT is not predicated on the same labor-technology relation.

For example, AT exhibits a myopic emphasis on manual labor rather than on the liberation potential of technology. Work is glorified as evidenced by this description of ten beggars employed through the use of AT:

They are completely rehabilitated now. They are well-dressed; they are happy; and are now completely different people. . . . It just shows that if one has the imagination, almost anything can be done (Lofthouse, 1977, p. 156).

Technology is cast in the role of rehabilitation, which translates as "put people to work." Unemployment, from this perspective, is taken as a consistently overriding factor in technological choice. But a technology freeing people of arduous tasks can be just as important as one creating jobs. After reading AT literature it is difficult to avoid the feeling that major labor-saving technologies are reserved for the First and not the Third World.

Further problems arise with the AT view of manual labor. For not only is there an emphasis on production, but even more narrowly there is a focus on the male productive functions of technology. Only the male aspects of technological development – innovation, design, construction, supply – are given attention. The other side of technology – daily operation, maintenance, use, care, responsibility – are ignored. An heroic image is fostered which finds little if any serious value in the nurturing functions. There is clearly a sexual division of labor between those features of technologies most commonly associated with males and those associated with females.

Devastating consequences, including the failure of many AT projects, result from the sexual division of technology in the AT movement. In the 1970s, windmills were sprouting up all over the northeastern part of the United States, including one atop a rehabilitated building in the New York City Loisada project. That was the exciting phase. Now few of these windmills are turning because, among other things, no one thought to maintain them. The dull, nurturing phase was bypassed. Likewise those technologies impacting primarily on women, such as reproduction, cooking, sanitation, and education, take a back seat to the gadgetry and engineering feats needed to build appropriate technologies.

Much of the critique just presented comes from Pacey's *The Culture of Technology* (1983). Yet, Pacey refuses to apply this critique to AT. I find nothing in AT practices to exonerate it from his charges.

It would appear that AT is not liberatory by design. Freeing people from work and addressing nurturing activities are not a central concern in the AT framework. So, labor intensity betrays something deeper about the AT movement than its concern to employ people. The AT movement has certainly been influenced by the environmental movement. Comparable influences, however, from the libertarian and feminist movements are not as evident or prominent.

C. *Small scale vs. indigenous technologies*

A convenient way to highlight the political bias underlying the small scale condition is to introduce another variant of AT, intermediate technology. Although there are differences between appropriate and intermediate technologies, these do not need to concern us here. The important thing is that intermediate technology stands politically in marked contrast to indigenous technology.

Commonsensically, intermediate means between low and high. The lower rung of the ladder – cheap, primitive, indigenous technology – is contrasted with the higher rung – expensive, mass-produced, foreign-based technology. Intermediate technology, recognizing "the economic boundaries and limitations of poverty" (Schumacher, 1973, p. 179), presumably lies somewhere between these two extremes. It is allegedly a progressive step above traditional technologies, and a step below high technology in that it is "production by the masses," not mass production.

So, when AT advocates proclaim that small is beautiful, the smallness condition is generally thought not applicable to indigenous technologies, despite the fact that these indigenous technologies generally satisfy all of the conditions. The reason for this is that despite Schumacher and despite some direct AT disclaimers to the contrary, the AT movement, in the final analysis, is capitalist. Generally, appropriate technologists focus on questions of productivity and expanding market structures. Production of useful commodities for the market is positively valued. Accordingly, crafting religious artifacts, which might encourage community cohesion and discourage marketable production, does not constitute an AT activity.

The AT bias against indigenous technologies is nicely illustrated by the following example of grain storage silos in Africa:

Many different types of traditional granary or silo exist, most of them built with mud walling. However, in some places there has been heavy loss of grain through the depradations of rats, insects, dampness and mould, and this has contributed to food shortages and malnutrition. Initially, it was assumed that such inefficiency was an inevitable part of the traditional technology, which was dismissed as almost worthless (Pacey, 1983, p. 151).

Because of the bias against traditional technologies the immediate action taken was to design concrete and metal alternatives to the indigenous ones. However, in this particular case AT innovations proved a failure, and attention was turned to the more nurturing task of suggesting "detailed improvements that would make maintenance easier" (Pacey, 1983, p. 151) on the traditional silos. Among AT practitioners this was a rare return to indigenous techniques.

Again, what is at stake here is not a choice between two kinds of technology, small scale appropriate or indigenous technologies. Rather contrasting philosophies about indigenous people are at stake. Paternalism in the form of "knowing what is best for the natives" is readily apparent among AT practitioners. AT proponents emphasize learning and adopting techniques which replace or notably upgrade a community's technical capacity. But the reverse is hardly ever the case: learning from and adopting indigenous technologies. This is most evident in the area of medicine. Western medicine does not hold a monopoly over healing. We could learn a great deal from indigenous medical theories and practices. Yet, as with almost all appropriate technologies, the important thing is not what we can learn from the indigenous people but what we can teach them.

Small scale is not an innocent condition of AT. Small scale, when conceived principally in terms of the production of marketable goods, not only yields a bias against indigenous technologies; it also hides an antipathy toward the powers and abilities of the very people it claims to help.

D. *Decentralization vs. radical technology*

Of the related versions – ecotechnology, liberatory, and indigenous technology – radical technology is preferable. This is not because it emphasizes a superior feature of technologies. In fact, it is just the opposite. Radical technology is least likely to be construed as only a philosophy about technology. On the surface, it makes little sense to search for the radical features of technological things. Radical technology is preferable because it is first and foremost a political philosophy which encompasses the political aspects of ecotechnology, liberatory, and indigenous technology.

At this stage I need to counter a possible objection. By isolating the AT conditions I seem guilty of the very charge I lodged against the critics of AT. This claim is easily met by noting that, unlike the case with replies to other critics, AT proponents in this case do not have the option of appealing to the virtues of one condition in order to offset the criticized vices of another. For example, the ecotechnologist's criticism of AT, that it only considers environmental problems in isolation, is not countered by appeal to labor intensity or any of the other conditions. Moreover, radical technologists try to overcome the problem of pitting one form of domination against another by searching for the common roots of domination.

Although some attention is paid to the importance of political considerations in making technological choices, politics is regarded by AT proponents as secondary to the technology itself. Indeed, technology seems to be conceived as a substitute for politics.

The choice of technology is the most important collective decision confronting any country. *It* is a choice that determines who works and who does not; that is, who gets income, new skills, self-reliance. *It* determines where work is done, whether concentrated in cities or more decentralized in smaller units; that is, *it* determines the kind of infrastructure required, and the whole quality of people's lives. *It* determines the ownership of industry; huge technologies are available only to the rich and powerful, whereas small technologies are tools in the hands of the poor (McRobie, 1981, p. 192; emphasis mine).

In ascribing these awesome powers to "it," McRobie's characterization of technological choice certainly has undertones of technological determinism: supply the Third World with the right kind of technology, and their lot will be improved. Form (engineering, technology) can be substituted for content (values, politics). Instead of the murky waters of values and politics people can turn to the relatively clear waters of engineering and technology. Seen in this light, it is not at all clear whether AT is an alternative to the technocracy esteemed within high technology circles; both appropriate technology and high technology advocates seem to be cut from the same cloth of "technology pushers."

Radical technologists reject the technocratic approach, proposing that political choices of what type of society to live in, not the choice of a technology, is "the most important collective decision confronting any country" (Harper and Boyle, 1976, p. 8). Technological choice is seen as first and foremost a political choice. Technology does not simply appear on the scene as an exogenous circumstance necessitating a particular social choice.

For the process of technological development is essentially social, and thus there is always a large measure of indeterminacy, of freedom, within it. Beyond the very real constraints of energy and matter exists a realm in which human thoughts and actions remain decisive. Therefore, technology does not necessitate. It merely consists of an evolving range of possibilities from which people choose (Noble, 1984, p. xiii).

The preoccupation of AT proponents with the qualities of technique ignores the existence of political power, of who decides and for whose benefit.

While AT philosophy seeks to avoid direct engagement of the problem of political power, this does not mean that it is without a specific political stance. AT proponents may shun political debates and feign political neutrality, but as I have tried to show there is a political philosophy and agenda underlying the AT movement. In general, by refusing to challenge and confront the political *status quo*, the politics of AT is largely one of accommodation and reform.

Some AT advocates interpret the decentralization condition in the explicitly political sense of "amenable to mastery and maintenance by local people" (Shaikh, 1981, p. 26). Nevertheless, closer examination of this interpretation shows that the idea of *control* by local people is entirely absent, and in AT practice local people seldom, if ever, take control of technological development. Decentralization in AT projects seems to be largely managerial and presupposes a technical hierarchy of AT innovators forming one class and AT implementers another. So, even where the political philosophy of AT is made relatively explicit, it no more actually challenges the political power structure than does high technology.

The aim of AT is to increase the number of technological choices, not necessarily to challenge any existing political structures.

What the [AT] proponents are trying to do is to open up the spectrum and find solutions which are better suited to local conditions. The aim is generally not to replace the existing industrial system – but to promote technological innovation in the areas where it has been, until now, either weak or ineffective (Jequier, 1976).

AT differs from high technology more in terms of the range of technological choices presented than in the array of political options debated.

Reformist politics coupled with the failure to take into account or challenge the political/social context of technical application has resulted in some AT projects becoming instruments of class domination. For example, in India only wealthy farmers in one region are able to make use of a methane gas plant (Shaikh, 1981). The two people able to afford a solar pump now monopolize the sale of previously free water in a Mauritanian village (Jequier, 1976). Small irrigation machines in Gao, Mali, enable farmers to sell enormous quantities of melons to Parisian tables, while food crops for local people are in short supply. (For further examples, see Lappé and Collins 1978.)

An example of the failure to account for political context but with a different result is provided by Elliott (1985). While working for an environmental group, he invented a hand-held radiation backscatter device which would measure the water content of a coal pile, resulting in more efficient utilization of coal. The device, however, proved highly unpopular among plant managers since their promotions depended on their coal burning efficiency rating. They could easily improve this rating on their own by reducing the water level in the coal. An "objective" measuring device would spoil this political maneuvering. In this case, an AT device's introduction was resisted because it would diminish the political power of management.

Without explicit non-reformist politics, the AT movement will always remain indifferent to political outcomes, and this exacerbates the problem of technological hierarchy and dominance. As a condition of AT, "decentralization is reduced to a mere technical stratagem for concealing hierarchy and domination" (Bookchin, 1980, p. 79).

E. *Advantages of radical technology*

I do not want to leave the impression that the radical technology proposal only gets its strength from its critique of AT. Radical technology is a rich and powerful political philosophy in its own right. Although obviously I do not have the space to provide a full rendering of its theory and practices, an outline of its basic tenets can be given.

How does radical technology combine ecotechnology, liberatory, and indigenous technologies as political philosophies? There are two main foundation stones supporting the radical technology framework. The first consists of a theoretical critique of and concerted opposition to domination. Domination comes in many different forms. Economic, political, gender, and race forms are a few of the more commonly recognized varieties of domination. Radical technology develops an analysis of each of these and their interrelationships. Moreover, radical technologists seek to reveal the connection between these forms of human domination and domination of nature: "The attempt to dominate nature stems from the domination of human by human" (Bookchin, 1980, pp. 66–67; Leiss, 1977). Any political philosophy choosing to ignore the domination of nature is doomed to foster the growth of domination structures rather than combatting them. So, radical technology combines ecotechnology and liberatory technology by providing a general critique of domination.

Empowerment, the second foundation stone of radical technology, means providing people with control over their institutions and practices. As we have seen AT practices can result in wresting technological control from people. The AT movement's major concession to popular control is a very weak interpretation of decentralization. In contrast empowerment is central to radical technology.

However, by promoting empowerment radical technologists do not thereby romanticize indigenous technologies and their accompanying political/social structures. Traditional practices can not only engender the domination of nature, they can also be forms of domination in themselves. Yet, my own hypothesis is that the traditional examples of domination count for little in the overall structure of domination. Nevertheless, the radical technologists should be aware of all forms of domination.

Radical technology (RT) is superior to appropriate technology because unlike the AT program the RT agenda is first and foremost an explicit political philosophy. Anything else evades the basic issues confronting any philosophy of technology. The issues are political and not technological in nature, where politics is defined as "any persistent pattern of human relationships that involve to a significant extent power, rule, or authority" (Dahl, 1970, p. 16). To see the issue of technological choice as not involving these relationships is to avoid seeing reality.

Moreover, RT is superior to AT in its political stance. Although AT proponents do not make a habit of making their political beliefs explicit, they do have them. Overall the AT political philosophy is to advocate very moderate piecemeal reform of the existing power structures. Yet, it is just these power structures which are impediments to a just, egalitarian, and democratic development of technology.

CONCLUSION

Contrary to Sardar and Rosser-Owen, AT is a coherent philosophy. It is not "a collection of a large number of ideas and concepts, many of them quite incoherent" (Sardar and Rosser-Owen, 1977, p. 564). A coherent program can be found in the prevailing ideas and practices of the AT movement.

Yet, the problem, by this analysis, is not the development of a coherent AT program. Attention to each definitional strategy showed that the AT movement has failed to develop its political side. Under the guise of insisting on ostensive definitions, a technocratic impatience with political questions was uncovered. Listing evidentiary features turned out to be an inappropriate way of giving political criticism, and qualities of techniques were elevated as political ends in themselves.

Finally, political problems were found even with the methodologically adequate set of conditions of AT that I proposed. With its environmental soundness condition AT, unlike ecotechnology, treats environmental problems as subsidiary, as isolated ones for which technical fixes can be found. AT proposals refuse to confront the broader, political problem of dominating nature. By emphasizing the labor-intensive quality of technology, AT is not liberatory in the sense of freeing people from arduous tasks nor in the sense of developing nurturing aspects of technology. The small scale condition is prone to an emphasis on marketable production and consequently devalues indigenous technologies and alternative economic forms and goals. The AT program as a whole exhibits a very muted and reformist sense of politics which neglects the political/social implications of its use.

Appropriate technology is often called alternative technology. To an extent, that claim is valid. High technologists and appropriate technologists deal with different technologies. However, it is delusional to think that AT offers very much of an alternative philosophy of technology – particularly when one factors in the political components of that philosophy. A truly alternative philosophy of technology requires the development of an ecological, liberatory, and thereby radical political foundation for technological choice. Philosophically, the first and most important goal is an adequate politics of technological choice.

University of Florida

REFERENCES

Bookchin, Murray. *Post-Scarcity Anarchism* (Palo Alto, Calif.: Ramparts Press, 1971).
Bookchin, Murray. *Toward An Ecological Society* (Montreal: Black Rose Books, 1980).
Bookchin, Murray. *The Ecology of Freedom* (Palo Alto, California: Chesire Books, 1982).
Brooks, Harvey. 'A Critique of the Concept of Appropriate Technology', in F. Long and A. Oleson, eds., *Appropriate Technology and Social Values* (Cambridge, Mass.: Ballinger, 1980).
Clarke, Robin. 'The Pressing Need for Alternative Technology', *Impact of Science on Society* **23**, no. 4 (1973).
Dahl, Robert. *After the Revolution* (New Haven, Conn.: Yale University Press, 1970).
Date, Anil. 'Understanding Appropriate Technology', in P. Ghosh, ed., *Appropriate Technology in Third World Development* (Westport, Conn.: Greenwood Press, 1984).
De Forest, Paul H. 'Technology Choice in the Context of Social Values – A Problem of Definition', in Long and Oleson, *Appropriate Technology and Social Values* (see above).
de Wilde, Tom. 'Some Social Criteria for Appropriate Technology', in R. Congdon, ed., *Introduction to Appropriate Technology* (Emmaus, Pa.: Rodale Press, 1977).
Dickson, David. *The Politics of Alternative Technology* (New York: Universe Books, 1974).
Diwan, Romesh K. and Dennis Livingston. *Alternative Development Strategies and Appropriate Technology* (New York: Pergamon Press, 1979).
Dunn, P. D. *Appropriate Technology* (New York: Schocken Books, 1979).
Eckhaus, Richard S. *Appropriate Technologies for Developing Countries* (Washington, D.C.: National Academy of Sciences, 1977).
Elliott, David. 'Energy Policy: Gut Reactions and Rationality', unpublished manuscript, 1985.
Harper, Peter and Godfrey Boyle, eds. *Radical Technology* (New York: Pantheon, 1976).
Jackson, Sarah. 'Economically Appropriate Technologies for Developing Countries: A Survey', in P. Ghosh, *Appropriate Technology in Third World Development* (see above).
Jequier, Nicholas. *Appropriate Technology: Problems and Promises* (Paris: Organisation for Economic Cooperation and Development, 1976).
Khalilzad, Zalmay, and Cheryl Bernard. 'Energy: No Quick Fix for a Permanent Crisis', *The Bulletin of the Atomic Scientists*, **36**, no. 10 (1980): 15–20.
Kochen, Manfred, and Karl W. Deutsch. *Decentralization* (Cambridge, Mass.: Oelgeschlager, Gunn & Hain, 1980).
Kohn, Leopold. 'Appropriate Technology', in S. Kumar, ed., *The Schumacher Lectures* (New York: Harper & Row, 1980).

Lappé, Frances Moore, and Joseph Collins. *Food First* (New York: Ballantine, 1978).

Leiss, William. *Domination of Nature* (Boston: Beacon, 1977).

Lofthouse, Paul R. 'Industrial Liaison', in Congdon, *Introduction to Appropriate Technology* (see above).

Lovins, Amory B. *Soft Energy Paths* (New York: Harper & Row, 1977).

McRobie, George. 'An Approach for Appropriate Technologists', in Congdon, *Introduction to Appropriate Technology* (see above).

McRobie, George. *Small Is Possible* (New York: Harper & Row, 1981).

Ndongho, W. W., and S. O. Anyang, 'The Concept of "Appropriate Technology"', *Monthly Review*, **32**, no. 9 (1981).

Noble, David. *Forces of Production: A Social History of Industrial Automation* (New York: Knopf, 1984).

Pacey, Arnold. *The Culture of Technology* (Oxford: Basil Blackwell, 1983).

Rybczynski, Witold. *Paper Heroes* (Garden City, N.Y.: Doubleday, 1980).

Sardar, Ziauddin, and Dawud G. Rosser-Owen. 'Science Policy and Developing Countries', in I. Spiegel-Rösing and D. Price, eds., *Science, Technology and Society* (Beverly Hills, Calif.: Sage, 1977).

Schumacher, E. F. *Small Is Beautiful* (New York: Harper & Row, 1973).

Shaikh, Rashid. 'Commentary: Reflections on AT in India', *Science for the People*, **13**, no. 2 (1981): 26.

Stewart, Frances. *Technology and Underdevelopment* (Boulder, Colo.: Westview Press, 1977).

van Brakel, J. 'Appropriate Technology: Facts and Values', in P. Durbin, ed., *Research in Philosophy & Technology*, vol. 3 (Greenwich, Conn.: JAI Press, 1980).

Winner, Langdon. *Autonomous Technology* (Cambridge, Mass.: MIT Press, 1977).

Winner, Langdon. 'Building a Better Mousetrap: Appropriate Technology as a Social Movement', in Long and Oleson, *Appropriate Technology and Social Values* (see above).

Winner, Langdon. 'The Political Philosophy of Alternative Technology', in A. Teich, ed., *Technology and the Future* (New York: St. Martin's Press, 1986).

DANIEL CÉRÉZUELLE

REFLECTIONS ON THE AUTONOMY OF TECHNOLOGY: BIOTECHNOLOGY, BIOETHICS, AND BEYOND

From the modern Western perspective, technical action promotes freedom because it allows for the satisfaction of human needs while favoring human initiative and guaranteeing human mastery. It thus also reinforces the moral status of human subjects. In fact, technology is usually conceived as instrumental and transitive action upon things. Through its functional, social, and existential neutrality, it opens the way for a human existence compatible with all the requirements of ethics – as Karl Marx presupposes when he predicts that once the industrial forces of production are provided with adequate relationships of production, "humanity will emerge from its pre-historical period." In the age of industrial abundance, the domination of one person over another will have no more reason to continue.

Optimism regarding technology presupposes belief in its instrumental or transitive character, in technology as involving action which emanates from a subject to be exercised on an object. In principle, such action leaves the actor untouched. But often this action is presented as mediated by the tool which guarantees, beside extra efficiency, the permanence and integrity of the human actor. By contrast, political power exercised by one human being on another can debase the one and corrupt the other. According to the classical scheme, political domination always runs the risk of transforming the person into a thing by denying moral initiative and integrity. In the process of mastering nature technologically, however, the person ought to remain intact in his conscience (with a sense of justice), preserving and even developing his status as autonomous will and end in itself. The implication of such an idea is that technology is in itself neither good nor bad. Purely instrumental, impersonal, totally subordinated to the will of which it is the mediator or the physical amplifier, it is the bearer of no ethical quality. Without any positive character of its own it can be used for good as well as for evil and it leaves the human being wholly free in relation to things, society, and everyone else.

Such is a brief characterization of the conception of technology which has been at the basis of the Promethean optimism of the West for a number of centuries. Although questionable, this traditional theory has

Paul T. Durbin (ed.) Technology and Contemporary Life, 129–144.
© *1988 by D. Reidel Publishing Company.*

been maintained in modern times because it suited quite well a bourgeois industrial society which rests on a merchant economy and the production of consumer goods. Consequently, for a long time one could look upon technical progress as an expression of human mastery over nature and things thanks to science.

But today, in association with science, technology equally permits an intentional, direct, and efficient action upon human beings. A quick, post-World-War-II inventory shows the proliferation of techniques of propaganda, public relations, social work, psychotherapy, biological manipulation (i.e., genetic engineering), behavior control, neurosurgery, communication, information, planning, land management, etc.

Thus, no longer only the subject and actor, the human being now becomes the object of his technologies. In fact, the science-technology association draws its strength from an indefinite capacity for objectification which, sooner or later, can no longer ignore humanity and society. Just as with nature, the domain of culture, in which humanity aspires to affirm identity and autonomy, becomes in its turn and thanks to the progress of scientific reason, susceptible to manipulation and to being modeled according to a logic which relates neither to the person nor to freedom.

The result is that the difference and distance between subject and object, which our philosophical tradition has always maintained as an obvious structure of action as well as the foundation of morality, insofar as it permits a person to judge effects, to criticize the means toward his ends – this distance can henceforth be abolished. Technology today has really gone from transitive to reflexive: human action is often directed toward the human and is no longer simply an instrumental mediation between persons and things. This is why the old philosophical cliché of the existential, ethical, and political neutrality of technology ought to be firmly rejected. Modern technology is not only an instrument which one can use as one wishes in order to modify the state of things. Developing in subtlety as in strength, it has repercussions upon the human order and modifies it independently of the particular intentions and reasons which govern at the moment when it is put into operation.

In recognition of this fact, contemporary authors often speak of *the autonomy of technology*. By this they mean that some very important effects of technology in reality escape individual and collective choices, and are the result of a functioning peculiar to technology itself, which responds in only a weak way to moral, cultural, or political preferences and decisions.

This notion of an autonomous technology is certainly upsetting because, if true, one can no longer rely on the traditional appeal to values
and political good will to master and orient our social life. If empirically
confirmed, this notion ought really to lead us to revise our idea of
progress and in particular our idea of the relationship between freedom
and rationality. Perhaps for this reason it is difficult to discuss such an
idea, since the importance of the imagined consequences obstructs the
debate. Often an excessive attachment to a progressive and rationalist
conception of humanity leads almost automatically to a rejection of the
idea of the autonomy of technology, because this "Frankensteinian
monster" would be *a priori* incompatible with human freedom. But
before entering into speculative debates, it seems important to examine
to what extent this idea takes account of the facts and what limits we
should assign to it. This is what I propose to do, using the biotechnologies whose rapid development in the course of the preceding decades
has raised passionate debates and important philosophical discussions.

FROM BIOTECHNOLOGIES TO BIOETHICS

It seems obvious that, henceforth, progress in biological technologies
effectively permits a profound action on the human, on everyone's way
of being, even on collective ways of being. This is why, in the face of
ever more subtle and efficient biological technologies, one finds reservations and concerns more and more frequently expressed. In fact, anyone
who engages power also engages the possibility of misuse, excess, even
abuse – and thus responsibility. As new powers are increasingly placed
at our disposal, we face new responsibilities. Without doubt, excess and
abuse have not been missing, but also cases of conscience and numerous
efforts, individual and collective, have encouraged attempts to define
the conditions for the proper use of biotechnologies.

From an early period medical doctors elaborated special ethical codes
for their profession, governing the use of their technical skills. The
Hippocratic Oath represents the traditional model, which has been
given an increasingly deontological form. With recent progress in the
life sciences, however, situations are created in which, on the one hand,
existing deontologies become insufficient, unadapted, and in need of
reformulation while, on the other, new groups of persons are entrusted,
often without preparation and sometimes without even realizing it, with
enormous powers over others. This is why for some fifteen years we
have seen a considerable effort being made to find a solution to the

multiple dilemmas posed by the development and utilization of those biological techniques which raise numerous social, ethical, philosophical, and political problems. Such has been the origin of bioethics, which attempts to define and codify limits on use and practice in biotechnologies insofar as they bear on human subjects.

As a discipline, bioethics emerged during the late 1960s and has developed rapidly since, facilitated by the sensitization of public opinion through well-publicized cases such as that of Karen Ann Quinlan. It was realized that without control and without new rules the development of biotechnologies, particularly in the medical field, risks unintended, contradictory effects with respect to both the individual and even the collective interest. This is why bioethics concentrates on the predictable consequences of the practice of each technology at the individual as well as at the social and political level; it examines problems which can result from it and attempts to define rules to follow and limits to respect. It is not necessary to enter into a detailed examination of this discipline and the particular problems which it involves; thousands of pages would be necessary – as is illustrated, for instance, by *The Encyclopedia of Bioethics* (1978) – to examine each problem, the way it is posed, and desirable responses. It is sufficient to enumerate the principal fields of application in this discipline: behavior control; death and dying; euthanasia; experiments on humans and the problem of informed consent; techniques of birth and reproduction; genetics; the dispensing of medical services; population or birth control; allocation of scarce medical resources; etc. Against such a background it is possible to venture some remarks of a general character.

It is certain that, faced with the sudden proliferation of the biomedical technologies, human beings must make an effort to civilize them, to reflect morally on the use of the new powers which they make available. The development of bioethics as a whole can be considered an effort to respond to an important problem – recognition that the use of biomedical technologies, their practice by technicians to resolve particular problems, is not sufficient to promote true progress. Restricted to a purely technical approach, one witnesses the multiplication of risks, misuses, unintended and even negative effects. This is the energetic claim of bioethics.

Note that if the use of biomedical technologies poses problems, it is not because they are used for deliberately evil ends. On the contrary, the good faith and intentions of those who put them into practice are often indisputable. The problem is that in themselves, due to their simple efficacy, these techniques have effects which go beyond the particular objective for which they were utilized in the course of a specific intervention. This is why bioethics is needed to define the rules for adjusting biomedical technologies to acceptable ends and in order to make sure that their practice does not engender effects contrary to those ends.

For example, as soon as resuscitation technologies enabled us to keep the sick alive in a state of extreme coma, a whole reflective process has been necessary to reach a clinical redefinition of death. This clinical redefinition of death was made necessary by at least three types of problems. In the first place, as soon as more and more perfect means of treatment are at our disposal an expanded definition of death seems necessary in order to guarantee the rights of each person to all the means available to keep him alive. As long as the ensemble of criteria are not met, a patient will continue to have the right to all treatments and services due the living. In the second place, progress in the means of resuscitation which permit us to prolong indefinitely, with technical devices, a certain number of physiological functions, even when all signs of conscious activity and hope of a return to consciousness has disappeared, calls forth a new definition of death – this time restrictive. The question now is of determining criteria which define not just the right to life, but the right to death. Once such and such signs have been observed, then technicians need no longer continue their interventions – they may turn off the respirator, unplug the heart monitor, etc., and let death take its course. But action here is complicated by, in the third place, new surgical and immunological techniques which permit vital organ transplants. There is increasing pressure to remove certain organs as soon as possible from the patient – abuse here consisting of not waiting until the "donor" is completely dead. The combination of these three dimensions renders the question of the definition of death extremely delicate and important, and an abundance of bioethical literature is dedicated to it. And numerous such delicate questions are raised by developments in biological science and technology.

One can only be gratified by the development of the discipline of bioethics. Nevertheless – and this is a point to be considered at length – the development of bioethics in itself is not enough to bring to a satisfactory resolution the problems posed by progress in the biomedical technologies. This insufficiency ought to be clearly stated, otherwise bioethics actually runs the paradoxical risk of playing a negative role. If one begins to consider it a wholly adequate and sufficient response to problems posed by biomedical technologies, this new moral discipline could actually obscure problems which it does not resolve. It could cloud the perception of such problems, depriving the will of determination to confront them, and therefore the possibility of finding a resolution to them. This limitation within bioethics thus deserves closer examination.

Bioethics represents an insufficient response because it is founded on an approach which is too narrow for the problems with which the progress in biological technology confronts us. The specialized literature of bioethics translates above all into a preoccupation with formulating moral rules, deontological codes intended to regulate the use of particular technologies in all biomedical fields. While progress in science and technology allows new practical applications, the bioethicists try to examine diverse problems posed by each technique, to formulate the principles which ought to govern it, as well as the limits which should be respected in order not to create wrongs to persons or society. This comes back to examining problems piecemeal one at a time as they arise. This process is too narrow both in regard to the way it apprehends the problems and in regard to the way it responds to them.

First, with respect to the apprehension of problems: As was pointed out, bioethics attends to problems discretely as new technologies create them. By the same token, this discipline is deficient in its ability to develop a more comprehensive analysis of the general impact of biotechnologies on humanity and society. Bioethics lacks a synthetic view which would permit it to grasp biotechnologies in their totality in relation to other aspects of social life. In particular, the relation between the development of biotechnologies and the total technical environment is not taken seriously into account. In this way one can accuse bioethics of looking at problems through the wrong end of a telescope and isolating them from their sociological and technological context. Doubtless it is necessary to respond to immediate emergencies, abuses, and misuses to the degree that they are confirmed, but this method entails a loss of practical efficacy.

Second, with respect to suggested responses, one observes a certain unrealistic character which assumes that it is only when one has a good analysis of the ensemble of characteristics of a problem situated in a comprehensive political framework that adequate piecemeal solutions can be proposed. Faced with problems grasped in isolation, bioethicists respond with particular recommendations aimed at isolated situations according to formal principles – in brief, with casuistry. Faced with the variety of situations created by the biomedical technologies, there is an attempt to formulate the moral principles which govern relations between persons. For example, in response to the technicians' power to prescribe psychotropic drugs, there is an effort to define the rights of patients or citizens, the rules which regulate what is or is not permitted. This centers attention on the relationships between persons, relations which are codified according to rights and duties. Ethical committees, even tribunals, can be charged, in cases of litigation, with making these rules respected.

The bioethical process is thus essentially juridical. Presupposing that the problems engendered by biotechnologies are the effects of voluntary interactions between persons, it tries to define the principles which ought to guide, even oblige, the will and intention of the interacting parties. This juridical application is itself founded upon a rationalistic and optimistic individualism which is only valid for persons conscious both of their actions and of their motives and having the latitude to act or not – that is, free, lucid, rational persons who, moreover, question themselves constantly about the value and purpose of their actions, responsible persons of whom society at large can demand accountability at any time. This moralistic and juridical individualism harks back to a second presupposition, namely, that biomedical technologies are practiced in a liberal, pluralistic society, in conformity with Anglo-Saxon democratic ideology, a society in which respect for the rights of persons is everyone's concern, permitting criticism, placing independent institutions of appeal and arbitration at the disposal of all. Finally, third, the bioethical process presupposes a series of conditions, namely, that biomedical technologies are practiced by easily identifiable agents, professionals working in an institutional context which is both clearly defined and transparent.

To the extent that these conditions can be met, the bioethical process assuredly allows the regulation of certain problems, the avoidance of abuses and excesses. To this extent alone can one be gratified by the development of bioethics, while remaining aware that many problems escape it. It is clear that the development of this moral casuistry largely abstracts from the concrete social conditions involved in biomedical technologies, social conditions which undermine the ultimate effective use of bioethics in the following ways.

First, this laudable effort to civilize biomedical technologies to produce only positive fruit conflicts with the fact that each technical advance multiplies the possibilities for applying other technologies. This general principle in technology is certainly applicable to the particular field of biomedical technology in which innovation proceeds at an accelerating rate. The laborious and delicate process of formulating deontological or juridical rules is constantly surpassed by different ventures in the technical field: the work of codification will have to be taken up again and again, so that to be effective and up to date bioethics must produce rules according to the rhythm of industrial production.

Second, biomedical technologies, in their progressive refinement, have an ever deeper and more powerful impact upon humanity while it takes a considerable time to civilize them because of their complexity and because their effects are not always easy to grasp and measure. It is to be expected that they will continue to advance at an increasing rate and that human risks will only increase.

Third, progress in bioethics is above all confined to medical technology. In this area, it is true, abuses and misuses can sometimes be corrected by institutionalized ethical reflection. For example, in the United States, the effort to redefine death clinically in order to control the removal of organs from the dying for transplantation to the ill who need them, has brought about the suppression of certain abuses. But by its own nature, the universality of technology ordains expansion in the use of biomedical technologies and their practice is not necessarily in the hands of professionals of a clearly defined status working in a transparent, institutional context. In fact, one can presume that the institutional conditions required for efficient control of the good use of biomedical technologies will cease to be effective from the moment that just anyone can put them into practice. Such is the price to be paid for the universality of science. Sooner or later the techniques become banal. They end up becoming accessible to all, with their use no longer effectively subject to control by deontological codes and juridical dispositions.

Finally it is not necessary to be a great sociologist to perceive that in contemporary societies the persons who practice the techniques are not in the situation of an ideal moral agent, totally independent. On the contrary, they are subject to organizational, professional, and ideological constraints which hinder their capacity for free examination and critical reflection about the use of those technologies at their disposal, because it is their profession, their livelihood, to put them into operation. It is a fundamental weakness of most ethical-philosophical discourse to fail to take into account this important aspect of the largely irrational character of human motivation.

In order to appreciate the breadth of the problems posed by biomedical technologies, however, it is essential to situate them in the most general context of technical development. For more than a century and a half the development of industrial technologies of matter and energy has not ceased posing different problems: social, cultural, political, and ecological, to the extent that we have not yet been able to master them. The same thing could happen with the human technologies whose emergence we observe today. Indeed, their rapid development and unanticipated side effects have the potential for affecting society with greater intensity than industrial technologies. In the context of a civilization already dominated by industrial technology, the development of human technology, especially biomedical technology, seems to have been inevitable, the more so since in many instances they seem to promise correctives for disorders and dysfunctions associated with industrial technology.

BEYOND BIOETHICS

It is not possible, then, in the face of the emergence of this new ensemble of technologies, to satisfy the requirements of ethics only by submitting particular practices to rules. Even if one does not wholly agree with their thought, it is indisputable that socialist reformers of the last century were correct in thinking that the development of industry, which was completely transforming society, could not be controlled by limiting themselves to the legal codification of relations between individuals. They understood that control of industrial technology calls for a social reorganization, and they tried to define in their way the broad outlines of a new political economy. Even today it must be realized that an approach to biomedical technology which is limited to codifying particular uses is insufficient, that it is necessary to engage the technical processes in their ensemble, which implies a new policy toward science and technology. For this it is necessary to proceed with a comprehensive analysis, to identify the impact of these technologies on humanity and society, and to situate that impact in the dynamics of modern industry and technology.

Now this is very difficult, because the effects are not of the same nature as the causes and it is necessary to discover them in their socio-political, cultural, and even spiritual dimensions, as well as in their ethico-juridical dimensions. The difficulty is further reinforced by the fact that most of the worrisome effects are due to a combination of diverse technologies.

In particular, it seems difficult to have confidence in the capacity of individuals to organize their actions according to rational ethical imperatives (for example, the principle of responsibility) when from now on it is the status of the individual himself which can be altered by the unexpected convergence of very different techniques. In fact, it seems that progress in biomedical technology makes possible an act in depth upon the subjective structures of individuality and identity which were up to now the inviolate base of his ethical capacity. Only a subject conscious of himself and of his identity can undertake to examine his world and to judge it. A person must first be sure of his own identity in order to want to act in a responsible manner.

But the traditional responses to the question "Who am I?" are going to become more and more problematic. In fact, the physical identity of the person currently rests on ever more fragile foundations. To say "I am this body" may become more and more arguable in the age of organ transplants, artificial limbs, bionics, sex changes. From now on the sense of uniqueness, of the fragile and irreplaceable character of the body, can only be weakened.

Even to answer "I am this consciousness," this subject which unifies thought, may become problematic. Techniques of behavior control can condition us in such and such a way; as we have known for some time through the experience of advertisements, our *desires* are not altogether our own. Our *emotions*, too, can be altered for others: our agressiveness and our sexual *impulses* can be controlled through drugs, by different methods of psychosurgery, or by cerebral stimulation.

Finally, it is going to become more and more delicate to answer with certainty "I am the son of x and y." In fact, a quick glance at the technology of manipulating heredity and reproduction readily leads to the following conclusions. First, with the perfection of new methods of human reproduction, it is no longer theoretically necessary for engendering an individual that one be born from the union of a parental couple, that one be carried in the womb nine months by a mother, etc. Moreover, with the techniques of cloning, the uniqueness of each individual is no more guaranteed than kinship. The fact that each one can refer to a parental origin as a point of anchorage for his personality and his condition can only become problematic. Yet uniqueness and kinship are two fundamental existential structures which support both the infinite value of each human being, different from all the others, and are the roots of continuity, physiologically (the union of sexes which engenders him) and psychologically (the certainty of having parents in law if not in fact). From now on modern consciousness is going to have to deal with the possibility and perhaps practical certainty that human identity across time and successive generations is subject to manipulation and transformations. But when humanity can be acted upon, transformed in body, thought, origin, and descent, then on what basis can a person presume to address his fate? Where is the subject of the ethical life? It is difficult to see the way out of a contradiction between capability and power, on the one hand, and freedom and responsibility, on the other.

This is why the recourse to ethical voluntarism (which consists of deducing obligations from a general principle of responsibility) is certainly a generous but insufficient attitude for regulating through their usage the impact of the biomedical technologies on human life and society.

It is even more ominous that many of the troubling effects of these techniques seem to escape deliberations and individual and collective decisions. In the frame of our technicist society, once these technologies are available, their operation and implications are largely independent of individual projections and desires, so that in practice one cannot separate good from bad effects. In fact, one always has both, for the use of a technique appears mainly like an anonymous and statistical phenomenon. When the means exist, they are used, and this use almost seems to be imposed of itself – which is, again, what is indicated by the idea of the *autonomy of technology*.

MEANING, LIMITS, AND ETHICAL-POLITICAL IMPLICATIONS OF THE AUTONOMY OF TECHNOLOGY

To speak of the autonomy of technology is to deny that technology is wholly neutral and subservient to human wishes. Once we have powerful and capable technologies we cannot place them in operation any way we want and exclusively for objectives we desire. Inevitably they bring effects which are not desired and are the cumulative global effect of a multiplicity of acts, each of which can be altogether legitimate, rational, and oriented toward an honorable end. This idea should in no way be inconceivable or mysterious, the process having become familiar to us through the analysis of economic phenomena. In practice, it seems illusory to imagine that we can always define the "good uses" of a technology, on the one hand, and voluntarily proscribe all the "bad uses," on the other; we must live with both. For although it is true that means are created with a view to some ends, from another angle they impose on us their own constraints. If this affirmation is correct for individual technologies, it is even more so when the question is one of an interaction between diverse technologies.

To be more precise, possibilities for control over our technologies are limited in two ways. First come tendencies to *self-regulation* or self-organization. There exists a *static* character which at a given moment influences technologies to organize themselves in a functional system. The interconnection and growing integration of technologies then determine to a large extent the ways in which a particular technology will be used. Use is determined not only by our will but also and above all by the cohesion of the technical context in which it occurs.

Next, possibilities for mastery and control are limited by the *dynamic* character of modern technology which seems to grow from itself. On the one hand, each innovation requires new ones to correct imbalances it creates. On the other, each innovation combines with technologies already in existence by widening their fields and possibilities of application. Technology seems thus to need its own growth and to exhibit a "snowball" effect. This growth seems like an anonymous process that no one directs. Everyone, even (and above all) those who do not "believe" in the autonomy of technics is convinced that the process will continue, whether we like it or not. Here again we run up against a contradiction between power and freedom which is not foreseen in classical ontology. Thus people never stop repeating the cliché that "you cannot stop progress." Optimism or fatalism? One can no longer distinguish between them as people "make a virtue of necessity."

Such then is what can be understood by the "autonomy" of technology. But the affirmation of this autonomy should be interpreted in a sociological not an ontological sense. It is not a question of a character necessarily attributable to any technology *sub specie aeternitatis*.

To begin with, the character of this autonomy is relative because it is only applied to that modern technology which is as of now inextricably associated with science and industry. Surely the possibility of a society in which technology would not exhibit such functional connections cannot be excluded. But the problem is that the more technology we have, the more difficult it will be to orient and control, because of the strength of its dynamism and the inertia tied to the integration of technical ensembles.

Next, the autonomy of technology must not be absolutized. For example, it would be illegitimate to say that technology has its own laws of development. The autonomy of technology is limited by the fact that it cannot function and, *a fortiori*, develop, without human acts. It cannot grow by itself; it is not a living organism. At the same time, everything seems to indicate that in our modern society technology is bound to grow and diversify. Although it does not grow except through human acts, dynamism and irreversibility nevertheless seem to constitute fundamental aspects of modern technology. And while it is true that technology cannot develop on its own without human acts, it must still be recognized that its progress as a whole is not directed as part of a program which was consciously elaborated by human agents. There is no pre-established plan. Most of the time technicians just try to put into operation innovations which were made possible by preceding innovations. No one can predict what technology will be like in a hundred years, because its development is not oriented toward some long term well-defined objective. It is not oriented toward an assignable telos. What technology is today is no basis for predicting what it will be in one hundred or five hundred years. The only thing that can be said is that it is headed toward exponential growth and capacity in means as well as an indefinite advance in internal coherence and in the functional integration of these technical means.

Thus, although human intervention is necessary during each moment of this growth, it seldom has a special direction that is significant to the technical ensemble. The use of each technology is determined largely by functional and social needs which are very often due to the already existing technical context. It is thus hardly reasonable to imagine that the operation of these techniques will engender only desired or desirable effects. Furthermore, once undesirable effects are manifest they become difficult to suppress because the technologies which are their cause are interrelated by a multitude of functional connections with the ensemble of technology. Now that computers are used everywhere, it is difficult to see how anyone can get along without them. It is only in this restrained and sociological sense that we can speak of the autonomy of technology.

At the same time, such a description implies that our situation is incompatible with the demands of ethics, since it overwhelms the very principle of human responsibility. As a result, it is no longer sufficient to define moral principles to regulate such and such a particular use of technology. Since the means themselves impose their own logic on social and personal life, the definition of new duties and principles does not constitute a wholly adequate and appropriate response to the problems. The priority should be instead to create a technical context compatible with an authentically human act, preserving the possibility of some effective responsibility. In this case, the problem is not so much knowing how to use such and such technique but deciding in the first place if it ought to be used. Today the crucial choice concerns not the principles which ought to orient technical practices and thus the subject's acts, but rather the very existence of things, the means of power over things and other human beings which should be the object of a choice. We should not reject *a priori* the idea that in certain cases it can be taken for granted that the only means of saving the practical possibility of an ethical life and freedom of action is to renounce the power which technology brings.

If such an analysis is correct, we must conclude that in the name of practical (or ethical) reason, we should plan to restrain the scientific-technical use of reason and accept the paradox of reason which turns against itself to limit itself, in order to maintain a technical context in which the opening to choice and responsible acts holds a central place. This is why our principal duty today is to choose between our technologies and remain this side of a threshhold of power and complexity beyond which we will no longer be in a position to control their use and social impact. Necessarily, the pre-condition for this choice is to stop this kind of automatic process of technical growth in which we live today. If not, we will never be in a position to take the time to think about what we are doing.

In fact, it is precisely because humanity is destined to live indefinitely with powerful scientific-technical forces and knowledge that the progress in development of technology must no longer be automatically identified with that of reason. Past a certain historic threshhold, do not the facts oblige us to admit a conflict between knowledge and morality? Doubtless the conflict does not appear if one remains only at the level of individual acts and deliberation; it appears only as one leaves the rationalist fiction and takes into account the concrete social context. If the sociological, economic, and technological determinants which in fact weigh on individuals are taken into account, it is difficult to see how we can pretend to regulate ethically the social use of ever more powerful technologies. Realism requires that we abandon this illusion and consider, instead, giving up that research which, for example, is likely to provide the means for altering the genetic identity of humanity, or for controlling individual behavior. Taking into account the irremediable character of human and social imperfection, it seems reasonable to limit knowledge and technology, at least provisionally, this side of that threshold beyond which we no longer control the effects.

Such a proposition is naturally repugnant to a culture which sees in the indefinite and unlimited pursuit of science and technology an expression of human greatness which nothing should impede. But two arguments support this proposal – one of simple prudence, the other from metaphysics.

The idea of a moratorium, that is, of putting a provisional brake on scientific-technological progress can be derived from a simple principle of prudence. No one would be mad enough to drive at great speed at the wheel of a powerful automobile without being assured that the brakes worked. The only way to be assured that the brakes work is to try them. We who are engaged in a process of accelerated change are in exactly the same situation. To succumb to an intoxication with speed only conceals the fear latent in the common saying "you can't stop progress" – a saying which at once testifies to an awareness of the autonomy of technology but wishes to insist on its beneficence.

A metaphysical argument supports the notion of a more definitive halt. Is it not in fact unreasonable to assume that in a finite world (such as our biosphere) technical progress will not engender dramatic effects? Sooner or later the expansion of human knowledge and power can only come into contradiction with the limits of our finite world. To argue the contrary amounts to declaring that humanity is capable of perfection – or can simply change planets as often as needed. Such an idea is more akin to bad theology than true philosophy. One can reply, of course, that such drastic limits have not yet been reached – although the uncontrolled proliferation of nuclear arms is surely evidence that in one area it certainly has. Nevertheless, it is difficult to deny that, sooner or later, they will be. If they are not chosen they may impose themselves in a very unpleasant way, which does not seem desirable and argues in favor of preventive self-limitation.

To conclude, note that recognition of the autonomy of technology in the limited sense described does not necessarily lead to an anti-technological "Luddite" stance. To say that we should impose limits on technology does not imply hostility to technology. Discussions on this matter should not be boxed into the false dilemma of all or nothing. It would be stupid to suppose that we must either accept all technology or else reject it totally, to go back to the candle and chariot, etc. To argue that we must not go beyond desirable boundaries of complexity and power is not to propose the renunciation of all technology – although we do not need all the techniques available today. Moreover, one of the essential elements undermining possibilities for the control of technology is the contemporary speed of innovation and its spread, which excludes evaluation and control.

Therefore, it can be projected that beyond a certain boundary the progress of scientific-technical reason will prove to be illegitimate – even when, within its own realm, it retains a certain flawless perfection. Such a proposition should not be unexpected. Did not Kant already show the need to limit the speculative uses of pure reason because of contradictions between the characteristic tendency of human reason (which imagines the infinite and the absolute) to defy limits, and the finiteness of our condition in which human experience must conform to a necessary spatio-temporal framework? To propose limiting the scientific-technical use of reason in the name of contradictions between the indefinite growth of power and the admitted finiteness of humanity and the world is to be situated in a comparable perspective, one which has yet to be rejected as inimical to logic.

Aquitaine Regional Institute for Social Work and Social Research

Translated by Frederic Courtney and Carl Mitcham
Philosophy & Technology Studies Center, Polytechnic University

DANIEL O. DAHLSTROM

LEBENSTECHNIK UND ESSEN:
TOWARD A TECHNOLOGICAL ETHICS AFTER HEIDEGGER

I

For some students of his writings, Martin Heidegger's alarming accounts of technology are difficult, if not impossible, to square with his denial of having anything against technology.[1] Heidegger himself fended off such criticism by insisting on a distinction between technology and its essence, "essence" not in the sense of a genus, but rather in the sense of something that presents itself as the pervading and perduring way in which things unfold. Already in the *Overturning of Metaphysics* he told his readers that he was not employing "Technik" for "the separate regions of mechanized production and equipment," but rather as a term encompassing "all the regions of particular beings which outfit the entirety of beings: the objectified nature, the managed culture, the manufactured politics, and the superstructure of ideals."[2] In *The Question Concerning Technology* the distinction between technology and its essence is made explicit: "What is dangerous is not technology. There is no technological demon, but there is a secret to the essence of technology. The danger is the essence of technology as a fate of revealing."[3] Thus, when Heidegger made his remark about being unopposed to technology, in his interview with Richard Wisser, it seems reasonable to assume that he was referring to technology in the popular sense of sophisticated implements, machinery, and processes involved in their use. Indeed, in another public setting, this time the memorial address published as *Gelassenheit*, his remarks are similar: "It would be short-sighted to want to condemn the technological world as the work of the devil. We are oriented to the technological objects; they even challenge us to an ever-mounting betterment."[4]

Nevertheless, Heidegger does exhibit a preference for certain sorts of technologies over others. Windmills and working the land by hand are at one point favorably contrasted with nature-challenging hydroelectric dams and motorized farming ("motorisierte Ernaehrungsindustrie").[5] The essence of technology itself, its fateful way of revealing the presence of things only by challenging and ordering them in some structure, is "the supreme danger" ("die höchste Gefahr") because it eliminates

145

Paul T. Durbin (ed.) Technology and Contemporary Life, 145–159.
© 1988 *by D. Reidel Publishing Company.*

any other way of revealing the essence of things, including the essence of mankind itself.[6] The history of metaphysics, that basic move or feature of Western-European history ("Grundzug der abendländisch-europäischen Geschichte"), is for Heidegger the history of the forgottenness of being – *Seinsvergessenheit*, the repeated reduction of being to some scheme of particular beings, and technology is the perfected or complete metaphysics ("die vollendete Metaphysik").[7] The charge of inconsistency, especially as it concerns the essence of modern technology, would seem to have some validity after all. Moreover, if this criticism can be sustained, then there is some legitimacy to the complaints that Heidegger's thinking is agrarian and aristocratic ("bäuerlich, aristokratisch antikisch"), naively romanticizing nineteenth-century village life, and that it ignores the positive, emancipatory character of modern technology.[8]

Not incidentally, these difficulties with Heidegger's thoughts on technology replay a basic tension in Heidegger's early thinking. Just as in *Being and Time* the fundamental ontological investigation of human existence (*Dasein*) could not unhinge itself from ontic considerations and practical presuppositions, so Heidegger's account of technology's essence is not free from certain technological prejudices. This feature hardly subverts Heidegger's insights into technology, any more than his acknowledged predilections in *Being and Time* defeat the analyses of that work. (Presuppositions do not make an argument or an interpretation – though it is generally true that an argument or interpretation is only as good as its presuppositions.) Heidegger made no pretense to a presuppositionless beginning. His method was to investigate the meaning of being as it historically presents (and absents) itself in the everyday encounters of being-in-the-world. His concern was not so much with the specific things filling up the world as with the context-of-their-significance, *viz.*, the presence and absence of a world as the advance and retreat of being itself. The critical issue is not whether he presupposed certain things about the world he has experienced, but whether (and to what extent) that world discloses itself in his interpretation of authentic and inauthentic existence.

The same sort of query has to be put to Heidegger's studies of technology. Does his interpretation of technology describe how the essence of technology exhibits itself? This question is obviously as tied to the issue of authentic technologies as the question of being in *Being and Time* was tied to the question of authentic existence. Do Heidegger's thoughts give us any indications as to appropriate technologies or any implications for living authentically with technology? Can they?

II

The ethical ring of such questions presents distinctive difficulties in coming to grips with Heidegger's thinking in general. Certainly Heidegger was adamant that he was no existentialist, that the investigation of human existence (*Dasein*) in *Being and Time* was directed by the question of being and not some tacit moral or humanistic aim, that such notions as care and conscience were to be understood ontologically, that is to say, as presupposed – not provided – by anthropological and religious conceptions.[9] Long after Heidegger had abandoned the transcendental (ontological) structure and language of *Being and Time* (as being inadequate for the completion of its project), he continued to insist on utterly eschewing metaphysics and the humanisms and ethics resting on its confusion of ontic and ontological orders.[10]

Yet, while trying to undo the misunderstanding brought about by existentialist and humanist interpretations of his work, Heidegger did not disavow the humane interest informing his thinking. Indeed, humanisms were rejected because they failed to appreciate the actual dignity of the human being ("die eigentliche Würde des Menschen"). "To this extent the thinking in *Being and Time* is opposed to humanism. But this opposition does not mean that such thinking takes sides against the human and advocates the inhumane, defending inhumanity and degradating the dignity of the human being. Humanism is opposed because it does not set the humanity of the human being high enough."[11] Thus, he added later that "the human being is not the master of beings. The human being is the shepherd of being."[12] Not the lord of beings, a human being possesses the simple dignity of the shepherd entrusted with the care of being.[13]

The shepherd, in Plato's image in the first book of the *Republic* (345d), illustrates that the being of what is shepherded defines the keeper of the flock. That this is humanism "in the extreme" and even "in itself already the original ethics" Heidegger readily acknowledged.[14]

Nor did Heidegger pretend that this "more original ethics" makes no political statement. Only by thinking the historical forgottenness of being that marks modern technology, Heidegger insisted, can we hope to understand the homelessness of contemporary humanity as well as the inhumanity of nationalism and collectivism alike. Communism is not merely a party or world view, and Americanism is not simply a particular life style; rather both are to be understood in terms of the forgottenness of being, the essence of technology, historically bestowed on us by being itself.

If he persisted in eschewing the concepts and terminology of traditional humanism, ethics, and practical philosophy, Heidegger's reasons for doing so were straightforward. The limited success of *Being and Time* in recalling the question of being, Heidegger emphasized, was hampered enough and even falsified, he came to realize, by its necessary recourse to the terms of the prevailing philosophy. Employing traditional terminology promotes the tendency to represent in the usual ways and thereby to distort what is not usually thought. Thus, Heidegger had recourse to metaphor when he referred to human beings dwelling in the house of being. Only by such devices could he hope to turn attention to this sort of truth excluded from consideration by traditional, conceptual thinking.[15] Protesting as "a great misunderstanding" the criticism that he forgot and neglected human beings in his preoccupation with being, Heidegger insisted: "For the question of being and the unpacking of this question precisely presuppose an interpretation of human existence, i.e., a determination of the essence of the human being. And the basic thought in my thinking is just this, that being, or better, the disclosability of being needs the human being and that, *vice versa*, the human is only a human insofar as he stands in the disclosability of being."[16]

Heidegger's humanistic concern did not take the form of a program for rearranging things and people, a set of directives for personal and communal living, or a plan for manipulating technologies. Strictly moral and ontic issues – the typical concerns of most students of technology – seemed dominated by a misguided, sometimes subtle, sometimes explicit drive for mastery of beings, a sure sign of the forgottenness of being. "As long as we represent technology as an instrument, we remain caught in the demand to master it. We drive right past the essence of technology."[17] Before we raise what seems like the only pressing concern, *viz.*, what should we do?, Heidegger admonishes us first to ponder the question: how should we think?[18] For Heidegger the central issue for humanity, its fate in the nuclear age, is tied to a question, the question of being (*die Seinsfrage*). A question is not a principle and it does not ground or establish other, subordinate principles of a more particular scope, whether a domain of theory or practice.

"This thinking is neither theoretical nor practical. . . ."[19] But then, does this mean that Heidegger's thoughts on technology cannot, after all, tell us anything about living authentically with technology? Is anything like a technological ethics to be gained from posing the question of being and its more original ethics?

Heidegger was hardly unaware of the hardness and the apparent hardheartedness of his attitude toward traditional ethics. When asked why, after *Being and Time*, he did not write an ethics, he acknowledged the legitimacy of the question, given both the desperateness of the human condition in the twentieth century and the significance of the first steps taken in *Being and Time* to ask the question of being as the essential question for man.

Where the essence of the human being is thought so essentially, namely solely on the basis of the question of the truth of being, whereby, however, the human being is nevertheless not elevated to the center of beings, the demand for some binding direction must grow as must the demand for rules which say how the human being, experienced by virtue of an exposedness to being, should fatefully live (*wie der aus der Ek-sistenz zum Sein erfahrene Mensch geschicklich leben soll*). The wish for an ethics urges to be fulfilled, all the more passionately as the acknowledged no less than the repressed helplessness of the human being mounts to incalculable levels. Every care must be devoted to the ethical bond, where the technological human being, having been handed over to the masses, can still be brought to some reliable constancy only through an organization of his planning and acting as a whole, in conformity with technology.[20]

Here almost a decade before his memorial address at Messkirch (*Gelassenheit*), Heidegger displayed his realistic appraisal of the need to accommodate technology and to do so ethically. Indeed, cultivating and securing existing ties among men, even if they serve simply to prevent a worsening of the human situation (Heidegger wrote this in 1947–1948), is an ethical demand of the highest order, he readily conceded. Nor did he fail to recognize that his non-ontological ethics, far from dismissing the question of *praxis*, actually sharpens it.

III

Yet precisely in his writings on technology Heidegger seems to many of his critics to have painted himself – and humanity along with him – into a tragic corner. The pastoral and domestic imagery of Heidegger's letter to his young French friend gives way to the mercantile and industrial terminology of "orders" and "machines," "stock," and "framework."[21]

Heidegger continued to insist that humans are not masters of beings, but the emphasis shifted from human shepherding of being to being's fated challenge to humanity in modern technology.[22] Heidegger's term for the essence of modern technology, "das Gestell" (for which "contraption," "system," "structure," "implacement," and even, most literally, "shelf," are arguably appropriate translations) designates nothing man-made. Nor does it signify a genus for everything technological or the complex or supply of parts and processes involved in technological endeavors. Rather *das Gestell* is being itself as the fateful presence-and-absence of things, a way of presenting things by challenging them, trapping them (*stellen*) into revealing themselves.[23]

What is the technological doing of humans is a reaction to an imposition by being itself, trapping them into disclosing and uncovering things in just this way. The danger, alluded to earlier, is among other things that human beings fail to see this trap as an historical imposition by being itself. In this forgottenness of being nothing seems so natural as to consider ourselves lords of the earth, to consider things as objects and ultimately, even less than objects, as so much stock, items on a computer print-out.[24]

This interpretation of modern technology's essence as the fated imposition of being itself has been decried as "pessimistic mysticism of being" and "highly stylized, ontological tragedy."[25] According to these critics, Heidegger's talk of the provoking, challenging character of modern technology's essence mystifies man's relationship with nature, as though it were an inevitable duel human beings are bound to lose. The danger that human beings increasingly reduce nature, including each other, to some enormous supply dump is uncontestable, but why, it is asked, must it be construed as a demand made by being itself?

Such criticisms are academic. They betray a failure to appreciate how modern technology's essence constitutes the meaning of being in our age, even in supposedly non-technological moments and endeavors, and that this meaning, unlike technology itself, is not a mere tool or means of human making and for human manipulating. Moreover, such criticisms are particularly inappropriate inasmuch as Heidegger saw, not tragedy, but cause for hope in the essence of modern technology. While recounting the threat of modern technology, pushing human beings, as it were, to the edge, Heidegger nonetheless insisted that "if we open ourselves explicitly to the essence of technology, we are taken up unexpectedly into a liberating demand (*Anspruch*)."[26] He held out the possibility of turning to "a more original revealing" in which human beings engage themselves more originally ("eher und mehr und stets anfänglicher") in the essence of what is hidden.[27] Hölderlin's phrase: "Yet where danger is, thrives also what saves" ("Wo aber Gefahr ist, wächst/Das Rettende auch") became the refrain of Heidegger's thinking about technology, articulating both the finitude and the possibilities of technology's essence, the limitations and the promise of the contraption (*das Gestell*).

Hölderlin's words are to be taken literally. Any saving can only emerge from the essence of technology itself. "The danger is what saves, insofar as it brings what saves from out of its own hidden, turned essence" ("Die Gefahr ist das Rettende, insofern sie aus ihrem verborgen kehrigen Wesen das Rettende bringt").[28] In other words, the turn must be taken by being itself, i.e., in that contraption that is the essence of modern technology.[29] But herein lies being's (or we might just as well say the world's or the system's) liberating claim on humanity, though it is a freedom of a more original order than human willing as such.[30] This freedom can perhaps be best described in more traditional language as the capacity to act in accordance with one's nature. Being, it must be remembered, unfolds both to and through human beings. They are not impotently delivered over to technology, though it is never simply a human instrument or means.[31] If a turn is taken by the essence of technology, human beings would be used, to be sure, but only in their corresponding to the turn, which is to say, in their freedom.[32]

The essence of technology, then, is as Heidegger himself acknowledged, "ambiguous in the extreme."[33] This fundamental ambiguity explains and even to an extent excuses inconsistencies critics have found in Heidegger's comments on technology, such as his claim to have nothing against technology, while plainly favoring certain (ontic) forms of technology and construing technology as "the perfected metaphysics," the (ontological) fate of *Seinsvergessenheit* and the greatest of dangers. Heidegger's admiring remarks about the windmill were not intended as an advocacy of a return to such technologies, but as a reminder of a non-challenging, non-(modern-)technological way being has historically presented itself. Moreover, Heidegger's embracing of Hölderlin's phrase ("Wo aber Gefahr ist, wächst/Das Rettende auch") confirms that the challenge to betterment modern technology presents to man is not to be understood in the first place as a demand to improve his lot in relation to other beings (ontically), but rather thoughtfully to prepare for a turn in the way being presents itself.

Yet, given Heidegger's own statement of the ambiguity in the essence of technology itself, the point of these criticisms can be formulated even more sharply. How can Heidegger consistently call for a stewardship of being when being itself presents this challenge to man? "Yet what is here called ambiguity, is it not still a contradiction of irreconcilable characteristics of the essence of technology? Can the same being with its unhiddenness on the one hand challenge the human being to plunge into the fury of ordering, on the other hand guarantee him in the same technology the fate of an authentic revealing?"[34] Furthermore, does not this description of the technological world force humanity into an attitude of unreserved resignation, "Lethargie" and "Immobilismus," as one critic has put it?[35] How else are we to understand Heidegger's remark: "We should do nothing but wait"?[36] And Heidegger's call for *Gelassenheit*, is this anything other than resignation, a barely disguised show of bitter defiance, suggesting that we "wait it out," as it were, like soldiers in trenches, until modern technology completely does itself in?

IV

An ambiguity is not the same as a contradiction, and it can only be banned in a perfectly rational world. Such a world, like a perfectly organized warehouse, is a fiction, though a fiction with unmistakeable power in our world today. Heidegger's enigmatic, often tortured prose is an attempt to make an unambiguous statement of the ambiguity at the essence of technology. The world displays itself in and through technologies that disguise the fact that the world is displaying itself. Instead, technology is construed anthropologically, as an expression of humanity, caught up in the disclosures, the production and organization, of things and objects in the world. Yet we know better; we know that the world unfolded by technology overdetermines our best efforts and interests; and we know this precisely because the essence of modern technology itself discloses it to us. Modern technology may be a deformity of life but it is still a life, a life we are able to live both because and in spite of modern technology. Such ambiguity holds both danger and promise.

The promise, like the danger, though not completely in our hands, does not come to pass independently of us.[37] "Composure" towards things, and not "resignation," best translates Heidegger's recommendation for living authentically with modern technology. Such composure demands that thinking, and not mere reckoning, be brought into play as a measure of things.[38] In other words, maintaining this equanimity toward things requires seeing them for what they are, a kind of thinking that does not pre-emptively subordinate things to special interests or to a particular scheme of organization. Heidegger was in dead earnest about the necessity of searching for a way out of the danger presented by modern technology, but the search can only begin with this composure toward things and the kind of thinking that sees the technological world for what it is, a world of danger and promise.[39] "Thinking we first learn what it is to live in the sphere where the overturning of being's fate, the overturning of the contraption, takes place."[40] In the terminology of Spinoza, we must learn to respond to *natura naturans* that composes and unfolds us.

Still, these remarks are obviously unsatisfactory. Heidegger did not presume to have secured a way out of the danger presented by modern technology. Indeed, he could not, any more than any other human individual or organization, inasmuch as the essence of modern technology is being or the world itself. Yet while the historical turn can only be taken by being itself, there are at least three clues to its promise. (There are other hints in addition to the ones singled out here, perhaps most notably, his accounts of dwelling and building and of art.) These same three signs may also be seen as ways of bringing into clearer focus what "composure toward things" and the liberation of such *Gelassenheit* means. Inasmuch as this specification of *Gelassenheit* sketches a human involvement in a more original disclosure of being in modern technology, it provides an outline for living authentically with technology, a *Lebenstechnik*, if you will. In other words, these signs are directional, suggesting the contours of a technological ethics after Heidegger.

The first mark of *Gelassenheit* is an appreciation of modern technology, not as a matter of theory or practice, but as the way being presents itself to us and through us. We might say that modern technology is "our way of life" provided we do not mistakenly see it as something we might acquire, master, and even discard. Indeed, human life itself is a technology, and not something outside us as an object or process to be used, admired, scorned, or abused. This technological life is best construed in the sense of "understanding" described by Heidegger in *Being and*

Time, viz., as a projecting of possibilities that is itself a disclosure of the world. Living authentically with technology, seeing it as the way being discloses itself, entails, Heidegger repeatedly insisted, that we grasp technology as more than an expression of human instinct or a means to human ends. Indeed, the Greek term *techne*, Heidegger insists, though translated in some contexts as "art," in others as "handicraft" (*Handwerk*), "never means a kind of practical accomplishment. The word means much more a kind of knowing."[41]

A second index to living authentically with technology specifies more precisely this non-theoretical and non-practical approach to technology. In characterizing the contemporary technological world Heidegger often cited the phenomenon that everything is made available to us, put on the shelves of our supermarkets and the screens of our TVs, and yet we are no closer to them. Great distances are overcome in milliseconds, yet nearness eludes us.[42] This loss of nearness is itself the progeny of *Seinsvergessenheit*, a metaphysical denial of mystery. Man is the neighbor to what is so near, *viz.*, being itself, that he hardly knows it is there. The nearness of being is language itself, the house of being ("das Haus des Seins") and man's housing ("Behausung des Menschenwesen").[43] That nearness of being prevails and yet escapes us – not *like*, but – *as* language itself. And certainly little effort is required to see how language is increasingly made into a tool. Words are cheap, word processors are easy, the creation of "language," for media and computers, is a business.

Yet language, like technology, is not the property, the expression, the tool of the *animal rationale*. No more than the being it houses is language essentially anthropological or psychological. Language can be objectifying but it need not be and in an important sense never is; that is to say, even when language is construed as merely a tool in the service of human aims and manipulation, it remains underdetermined by the human speaker or reader. The insight was Hegel's long before Heidegger penned the enigmatic phrase, "language speaks" (*Die Sprache spricht*): we always say more than we mean. Essentially, language is a corresponding to being. "Language (*Sprache*) is the original dimension within which the essence of the human being in general is first able to correspond (*entsprechen*) to being and its claim (*Anspruch*) and in the corresponding to belong to being."[44] Poets, with their ears and eyes for language, often are the better teachers here, as Aristotle once remarked, and Heidegger was fond of citing in this regard how *techne* itself was a form of *poiesis* in ancient Greece.[45] This poetics of language, in which language itself and not the poet essentially does the talking (*die Sprache spricht*), exemplifies not primarily the illusory technology of personal expression or social control, but a technology in which the world is able to disclose itself.

A final clue to a technological ethics after Heidegger can be garnered from his discussion of "Ding" and "Welt." In the system of the *Gestell* things are presented not in the act of being what they are and not even as objects confronting human agents and perceivers, but only as so much stock (*Bestand*).[46] More accurately, there are practically (i.e, in practice and almost) no more "things," since we face only commodities. The world becomes just so much productive capacity, permanently on order.

These hints are slight, as Heidegger himself acknowledged.[47] He made no presumptions about doing anything more than preparing the way for a co-response to a turn that must occur, beyond human control, in the essence of modern technology itself.[48] Yet the turn is not without human beings[49] and therein lie the possibilities for developing a technological ethics, a *Lebenstechnik*.

V

Can anything more positive be said about this *Lebenstechnik*? Can Heidegger's hints be developed more seriously and more precisely? Can various *Lebenstechniken* be identified and even specified in terms of these clues to a technological ethics?

Human beings exist only insofar as they act; their actions are what they are – but these actions are technologies. To distinguish, following Aristotle, making (or *poiesis*) as the mark of technology from being and doing, is simply inappropriate. Part of the rhythms of things, we maintain ourselves in various technologies from breathing, ingesting and digesting, sensing, and ailing to reading and love-making and contemplating. We act out, we enjoy, we suffer our technologies. Although each may take a form that kills us, they make our lives possible and hence can be considered *Lebenstechniken*. "Lebenstechnik" refers to that disclosure and unfolding – and the closure and folding up – of and by being itself that is the essence of modern technology.

Eating and food might be taken to constitute one such *Lebenstechnik*, to be elaborated in terms of the clues to a technological ethics, suggested by Heidegger's thoughts on language and things. Food can be and often is seen as merely a means and eating as just a technique, yet every morsel we consume has implications for us far surpassing anything we might have intended by placing it in our mouths. The Eleusinian mystery is the mystery of being, as near to us as language and as little comprehended. Nevertheless, we need not be told that the failure of this *Lebenstechnik* is our undoing and our unmaking. In eating we are caring for being itself and this is reducible neither to foodstuff and its preparation nor to mastication and the digestive tract. Eating as a *Lebenstechnik* is a world into which we have been thrown and a world we constitute. Indeed, this *Lebenstechnik*, the meaning of being disclosed therein, essentially remakes the earth and rearranges the environment.

These loose remarks can be made more concrete. Bread, for example, is no more simply a means of nourishment and enjoyment than it is simply a product of the food industry. Certainly the farmer and the elements, the miller and the market, the cook and the kitchen all must contribute to producing bread. But the bread is not what it is because it was produced; rather, it was produced because it is food. Nor is bread a complex of properties, as are designated by nutritional science. The triteness of these remarks is belied by the fact that manufacturers and scientists claim to assure a believing public what food is.

Food is on the shelves of our supermarkets today, bagged, boxed, jarred, and crated. Hardly an expense is spared in packaging and marketing in such a way as to make the food as attractive as possible or, at least, more attractive than the competition. Freshness and uniform quality are guaranteed, that is to say, forced out of nature by the power of competition and by a collusion of chemists and truckers – or your money back. All this marketing and packaging ingenuity, this organizational expertise in efficient production and distribution, all this commercial and chemical artistry – can they be said to disclose that delectable balance in which human beings live from the fruits of the earth? Why are "junk food" and "fast food" not merely ridiculous contradictions, but increasingly staples of the modern diet? Surpluses abound as farmers in Iowa are paid to let their fields lie fallow and European farmers destroy tons of eggs while Africans are dying of starvation. Because food is not seen as food, the earth's gift of nourishment to its species, but rather as the product of a competitive food industry, itself a tool and means of geopolitical scrambles, famine and abundance stare helplessly across a conference table at one another.

Little imagination is required to see what Heidegger's account of the essence of modern technology, *das Gestell* as being itself, threatening the reduction of everything to so much stock (*Bestand*), has to do with this modern technology of food and consumption. That being presents itself "everywhere" as something on order is not simply a way of looking at things, or even a way of producing things. The *Gestell* transforms food, like things and objects, into so much stock, something that can be put on the shelf. Here Heidegger's term "Gestell" is best translated literally. Being, like bags of flour and boxes of cereal, jars of peanut butter and crates of oranges, means "what-can-be-put-on-the-shelf" (*Gestell*).

The business of displaying grocery products in supermarkets and the marketing of menus in the mass media seem to force the presence of food upon us, yet the act of being food, its fragile presence and absence is obscured. Food is not an object, certainly not the theoretical object of the sciences of nutrition. Nor is food simply a thing, like a tree or the pitcher with a mysterious self-sufficiency. Food is what it is only in relation to an eating. Nonetheless, it remains something essentially other than a mere product or tool for humanity. The essence of modern technologies of food and consumption, moreover, constitutes a way being unfolds and discloses itself, that is to say, a way of nourishing and surviving, however much it also obscures this fundamental event. That event is a ritual celebration (eating) of a gift (food) of the earth and the sun to the human species.

The Catholic University of America

NOTES

[1] See Winfried Frantzen, *Von der Existenzialontologie zur Seinsgeschichte: Eine Untersuchung über die Entwicklung der Philosophie Martin Heideggers* (Meisenheim am Glan: Anton Hain, 1975), pp. 131–133; and Simon Moser "Kritik der traditionellen Technikphilosophie," in *Techne / Technik / Technologie* (Munich: Pullach, 1973), pp. 68–70; Richard Wisser, ed., *Martin Heidegger im Gespräch* (Freiburg/Munich: Karl Alber, 1970), p. 73.

[2] Martin Heidegger, *Vorträge und Aufsätze* (Pfullingen: Neske, 1978), p. 76 (hereafter referred to as "VA") and Martin Heidegger, *Die Technik und die Kehre* (Pfullingen: Neske, 1962), p. 20 (hereafter referred to as "TK").

[3] *TK*, pp. 27–28, 25.

[4] Martin Heidegger, *Gelassenheit* (Pfullingen: Neske, 1959), p. 22 (hereafter referred to as "Gelassenheit").

[5] *TK*, pp. 14–15.

[6] *TK*, pp. 26–28.

[7] *VA*, pp. 73, 76.

[8] See Moser, *op. cit.*, p. 69; Frantzen, *op. cit.*, p. 66.

[9] Martin Heidegger, *Wegmarken* (Frankfurt am Main: Klostermann, 1978) (hereafter referred to as "Wegmarken"), p. 313: "Auch die Namen wie 'Logik', 'Ethik', 'Physik', kommen erst auf, sobald das ursprüngliche Denken zu Ende geht." *Wegmarken*, p. 315: "Was in 'Sein und Zeit' (1927), par. 27 und 35 über das 'man' gesagt ist, soll keineswegs nur einen beiläufigen Beitrag zur Soziologie liefern. Gleichwenig meint das 'man' nur das ethisch-existentiell verstandene Gegenbild zum Selbstsein der Person. Das Gesagte enthält vielmehr den aus der Frage nach der Wahrheit des Seins gedachten Hinweis auf die anfängliche Zugehörigkeit des Wortes zum Sein." *Wegmarken*, p. 319: "Jeder Humanismus gründet entweder in einer Metaphysik oder er macht sich selbst zum Grund einer solchen." *Wegmarken*, p. 326: "Aber der Hauptsatz des 'Existentialismus' hat mit jenem Satz in 'Sein und Zeit' nicht das geringste gemeinsam; . . ."

[10] *Wegmarken*, pp. 319, 325.

[11] *Wegmarken*, 327: "Insofern ist das Denken in 'Sein und Zeit' gegen den Humanismus. Aber dieser Gegensatz bedeutet nicht, daß sich solches Denken auf die Gegenseite des Humanen schlüge und das Inhumane befürworte, die Unmenschlichkeit verteidige und die Würde des Menschen herabsetze. Gegen den Humanismus wird gedacht, weil er die Humanitas des Menschen nicht hoch genug ansetzt." See also *TK*, p. 32.

[12] *Wegmarken*, p. 338; *TK*, p. 41.

[13] *Wegmarken*, p. 339: "Sie (Ek-sistenz) ist die Wächterschaft, das heisst die Sorge für das Sein." *TK*, p. 41: "Nur wenn der Mensch als der Hirt des Seins der Wahrheit des Seins wartet, kann er eine Ankunft des Seinsgeschickes erwarten, ohne in das blosse Wissenwollen zu verfallen."

[14] *Wegmarken*, pp. 339, 353.

[15] See Heidegger's description of "die Technik als die höchste Form der rationalen Bewusstheit" in *VA*, p. 83, as well as his contrast of "das rechnende Denken und das besinnliche Nachdenken" in *Gelassenheit*, p. 13.

[16] Wisser, *op. cit.*, p. 69.

[17] *TK*, p. 32. Like Marx, most technologists want to change the world and not merely interpret it. But, as Heidegger pointed out, changing the world, far from its denial, presupposes an interpretation of the world; see Wisser, *op. cit.*, pp. 67–68.

[18] *TK*, p. 40.

[19] *Wegmarken*, p. 354.

[20] *Wegmarken*, p. 349.

[21] Compare *Brief über den Humanismus* (to Jean Beaufret) of 1947 with the four addresses of 1949 delivered under the title: "Einblick in das was ist."

[22] *TK*, p. 18: "Wenn also der Mensch forschend, betrachtend der Natur als einem Bezirk seines Vorstellens nachstellt, dann ist er bereits von einer Weise der Entbergung beansprucht, die ihn herausfordert, die Natur als einen Gegenstand der Forschung anzugeben, bis auch der Gegenstand in das Gegenstandlose des Bestandes verschwindet."

[23] *TK*, pp. 20, 23, 29, 38.

[24] *TK*, pp. 26–27.

[25] See Frantzen, *op. cit.*, pp. 131–133 and Moser, *op. cit.*, p. 67, 76.

[26] *TK*, pp. 25–26.

[27] *TK*, pp. 25, 28.

[28] *TK*, p. 41.

[29] *TK*, p. 44.

[30] *TK*, pp. 24–25.

[31] *TK*, p. 37.

[32] *TK*, pp. 39, 24–25.

[33] *TK*, p. 33.

[34] Moser, *op. cit.*, p. 79: "Aber was hier Zweideutigkeit genannt wird, ist es nicht doch ein Widerspruch unverträglicher Merkmale des Wesens der Technik? Kann desselbe Sein mit seiner Unverborgenheit einerseits den Menschen in das Rasende des Bestellens herausfordern, andererseits ihm in derselben Technik das Geschick eines echten Entbergens gewähren?"

[35] Frantzen, *op. cit.*, pp. 137–139.

[36] *Gelassenheit*, p. 35.

[37] *VA*, p. 174; *TK*, pp. 38–39.

[38] *Gelassenheit*, p. 21; *VA*, p. 174.

[39] *Gelassenheit*, p. 26.

[40] *TK*, p. 40: "Denkend lernen wir erst das Wohnen in dem Bereich, in dem sich die Verwindung des Seinsgeschicks, die Verwindung des Gestells, ereignet."

[41] Martin Heidegger, "Ursprung des Kunstwerkes" in *Holzwege* (Frankfurt am Main: Klostermann, 1972), p. 47. Hank Aaron was able to hit homeruns, but presumably unable to teach the art, to put it into some determinate theoretical form. Nevertheless, we say with perfect clarity that he knows how to hit homeruns.

[42] *VA*, pp. 157–158.

[43] *Wegmarken*, p. 330.

[44] *TK*, p. 40.

[45] *TK*, p. 44.

[46] *VA*, p. 174: "Im Ausbleiben der Nähe bleibt das Ding in dem gesagten Sinne als Ding vernichtet." In his unpublished essay, "die Gefahr," Heidegger noted: "Im Wesen des Gestells ereignet sich die Verwahrlosung des Dinges als Ding" (p. 18). With the neglect of the thing-like character of things, the world also retreats from view; the world illuminates and preserves the thing-like character of things, and is in fact the source of being itself, according to this manuscript. *Seinsvergessenheit*, the tradition of metaphysics come to maturity in modern technology, is rooted in *Weltvergessenheit*.

[47] *VA*, pp. 174–175.

[48] *TK*, p. 38.

[49] *VA*, p. 174: "Wann und wie kommen Dinge als Dinge? Sie kommen nicht durch die Machenschaft des Menschen. Sie kommen aber auch nicht ohne die Wachsamkeit der Sterblichen."

LARRY HICKMAN

THE PHENOMENOLOGY OF THE QUOTIDIAN ARTIFACT

In chapter four of *The Human Condition*[1] Hannah Arendt suggests that quotidian artifacts, such as the tables and chairs that we utilize on a daily basis, serve to stabilize human life. Between the private vagaries (one might even say the randomness) of human subjectivity on the one hand and the "sublime indifference of untouched nature" on the other, there is a man-made world protecting us from both.

She continues:

Only we who have erected the objectivity of a world of our own from what nature gives us, who have built it into the environment of nature so that we are protected from her, can look upon nature as something "objective." Without a world between men and nature, there is eternal movement, but no objectivity.[2]

Arendt ultimately concludes that it is neither in the activities of *animal laborans*, whose goal is to break out of his servitude to nature and the earth, nor in those of *homo faber*, who is both creator of human artifice and, consequently, destroyer of nature, that we find the measure of all use-things. Rather, it is the activities of man the thinker and doer that provide such meaning. Her and his activities have no aim outside themselves. They allow for a continually receding horizon of human dreams, hopes, and self-definition. But man the thinker can neither succeed nor survive without *homo faber*, that is, without "the help of the artist, of poets and historiographers, of monument-builders or writers, because without them the only product of their activity, the story they enact and tell, would not survive at all."[3]

The role of *homo faber* for Arendt is thus central to human life. It completes and supplants the activity of *animal laborans*, releasing him and her from their onerous tasks and stabilizing human life against the uncaring cycles of the household of nature. But it also renders palpable and permanent the various efforts of thinking man, rescuing him and it from the subjectivity of the private and unexpressed.

To those of us who hold the view that the humanities and the social sciences have too long neglected the implications of the concrete moments of human experience during their long and severe bondage to an almost exclusive preoccupation with the abstract and transcendent

161

Paul T. Durbin (ed.), Technology and Contemporary Life, 161–176.
© 1988 *by D. Reidel Publishing Company.*

features of our lives, Arendt's remarks are both suggestive and welcome. Her attention to concrete artifacts acknowledges and further
excavates the very touchstone which has been lacking in large segments
of the various traditions of abstract philosophy.

Arendt is certainly correct in arguing that both nature and human
subjectivity are active destabilizing agents with respect to the human
self. She draws on the traditions of the Greeks for this characterization.
Rationality, vested in the life of the *polis* (she is careful to distinguish
between the life of the *polis*, man as *zoon politikon*, and the life of the
family, *homo socialis*), was for them a bulwark against the irrationality
of solitary life outside the community, as well as a means of cooperation
against the debilitating and devastating forces of nature.

But some have argued, and I think for the most part correctly, that
our time has seen the advent of a "second nature": the rise of a complex
of technical methods and organization so extensive and pervasive that
contemporary human beings relate to it as their ancestors once did only
to the "original" nature. It, like nature, has its own inherent cycles and
laws of development. It, like nature, is to a great extent outside the
control of humankind. It, like nature, exhibits sequences of evolution
that are imperfectly understood even by those who spend their lives in
its study. Perhaps more than anyone else, Jacques Ellul[4] has documented the rise of this surrogate nature. He has also understood and
communicated to us many of the important ways in which it supplants
and destabilizes human freedom, just as the "first" nature before it had
done.

It is not necessary to buy Ellul's entire complex of arguments with
respect to this second nature. His critics are many, and much of what
they say has found its mark. Nevertheless, whether or not we wish to
speak of "technique," as he calls his version of this second nature, there
do seem to be large areas on the fringes of the consciousness of
technological man that appear to him to be beyond his control and
which threaten him. Such areas are marked, as Ellul has correctly
suggested, by their irreversibility, by their progression, which is geomet-
rical instead of arithmetical, by their mystery, and by the loss of sense of
scale and agency which they bring. These areas of awareness offer
resistance in much the same way that nature once thwarted human
initiative and accomplishment.

Expanding Arendt's thesis, then, let us say that the activities and
products of *homo faber* serve as instruments of stabilization against

subjectivity (and against nature, to be sure), but also against what I shall call, with intentional ambiguity, "Big Technology" – namely, those large, amorphous, and threatening aspects of technology that loom at the edges of our ability to understand and control our lives. But how does such stabilization take place?

Arendt has pointed the way to an answer with her suggestion that attention be turned to the functions of technical *quotidiana*. One such study was undertaken by Marshall McLuhan,[5] who examined media as the extensions of man and laid bare what he called "the folklore of industrial man." One of McLuhan's best known theses was that changes in technological systems or paradigms alter the most fundamental ways in which human beings interact with their world; and that the agents of these changes are the quotidian artifacts that most of us ignore because of our very familiarity with them. Another important analysis was undertaken by Edward Hall,[6] who did detailed cross-cultural analyses of the ways in which people use such humble objects as telephones and doors. More recently, Csikszentmihalyi and Rochberg-Halton[7] have undertaken careful analyses of the role of one particular type of techno-logical quotidiana, domestic objects, in the formation of the self. Their work is especially important in the sense that they seek to continue the work of the American Pragmatists, C. S. Peirce, John Dewey, and G. H. Mead, whose work has never been thoroughly mined for its many insights into this area. In true pragmatic fashion they are, for example, more interested in the *terminus ad quem* than in the *terminus a quo* of objects, and even analyze domestic objects as special kinds of signs, in the Peircean senses of that term.

But it seems to me that another rich source of insights into the functions of quotidian artifacts with respect to self-stabilization has been largely ignored by philosophers and social scientists. There is possible a double move to the support that Arendt thinks available from *homo faber* if we take the trouble to examine the works of mainstream fiction, itself a form of concrete making, for insights into the nature and function of quotidian artifacts, which are *homo faber*'s most basic products. I suggest that a good place to begin is with the novels of John Updike, perhaps the most "quotidian" of American writers.

I shall isolate and explore five modes of self-stabilization through technological quotidiana in Updike's work. I do not claim for these modes that they are either mutually exclusive or exhaustive. Neverthe-less, I believe that they shed light on Arendt's thesis that human beings stabilize themselves by means of the quotidiana in which they are enmeshed.

1. *Personation*: Technological quotidiana allow us to "personate" ourselves. I use the verb "to personate" in the sense of "to play at being oneself." Its correlative is "to impersonate," to play at being another. Douglas Browning, whose essay, 'Some Meanings of Automobiles,' explores this mode of self-stabilization, suggests that nowhere do we have more opportunities to personate ourselves than in our automobiled lives. Alone in the automobile we can sing in ways that would be unthinkable even in the shower. Such personation springs from a feature of the automobile that he calls its "anonymous privacy." The anonymous privacy of the automobile constitutes its paradox of inner and outer. I am alone in the privacy of my automobile. In the rush hour, I am surrounded by others, whom I can see, who are alone in the privacy of their automobiles. They, and I, feel free to do things we would never do in polite company. As Joseph Coates has put it, "If you are stopped at a stop light and look around, my observation is that about 10% of the people will be picking their noses."[8] Browning further characterizes this paradox:

Thus there is an inside and an outside to the automobiled human being. From the outside there is anonymity. From the inside there is privacy. The windshield is the frontier between the person and the world. The world is the place of exigencies, the coiled potentialities of violence, the possibilities of the sudden. The person is the autonomous master manipulator with the enriched intimacies of a mobile hideout.[9]

He suggests that love-play in the automobile, the mobile rendezvous, offers one of the richest examples of such anonymous privacy.

This is a mobile rendezvous. You enter myself as something stolen from the world. The secret is deep indeed. We may not linger, for the anonymity of our place is subject to immediate discovery. Wariness is mixed with the recklessness of intimacy. The outside peers into the inside. The ambiguity of sex is unique and exciting. By invitation the alien world admits to my trespass. A piece of the landscape transgresses my soul. I am myself the one who is most naked and vulnerable. I am at the same time the anonymous manipulator. Desperate passion is one with infinite reserve. The suddenness of the world wraps into our privacies, as though the automobile were turned inside out.[10]

Harry Angstrom, the protagonist of the "Rabbit" cycle, a character whom Updike has had monitor and mirror the national culture of the United States over three decades, is thoroughly enmeshed in quotidian apparatus. When we first meet him in *Rabbit Run* (1960),[11] he is a salesman for the MagiPeel Peeler Company, a task which he views as involving a kind of innocent fraud. It is his daily task to convince housewives that the MagiPeel will economize the vitamins in their fresh foods, thus putting their household economy on a firmer footing. After a stint as a gardener, which he compares to heaven, he goes to work as a used car salesman. In *Rabbit Redux* (1971),[12] Rabbit has taken up the trade of his father, who has been a linotype operator for twenty-five years. He operates in the contradictory world of machines that contaminate their operators with ink and grease, but create the order and uniformity of print. Both he and his father are eventually laid off, victims of the new electronic publishing technologies. In *Rabbit Is Rich* (1981),[13] he is a major partner and manager of a Toyota dealership, and this in the midst of the energy crisis of the 1970s.

Rabbit, like the rest of us, must come to grips with the alienated and alienating aspects of technology. But his refuge does not lie in religion, a possibility explored and rejected in *Rabbit Run*. Neither does it lie in an alignment with the natural processes of growing things. He cannot continue to be a gardener because of what he perceives to be his responsibilities to his family and to the wider demands of living in a technological society. Nor does his refuge lie in the romanticized naturalism of the counter-culture of the 1960s. His need for order, and his linotypist's and auto dealer's sensitivity to the human symbiosis with the machine react angrily against the oblivious neglect of the mechanical world, or else the outright Luddite activities, of that decade's flower children.

Rabbit's refuge is in quotidian apparatus. He uses a variety of them to personate himself. To the young Rabbit, whose problems with money, his wife, his job, and his self esteem seem to us insurmountable, the automobile is an occasion for self-discovery through play. Updike tells us that for him, "The car smells secure: rubber and dust and painted metal hot in the sun. A sheath for the knife of himself. He cuts through the Sunday stunned town."[14] Twenty-five years later, the response of Rabbit the successful Toyota dealer, the member of the Kiwanis and Rotary clubs and the Chamber of Commerce is much the same. Whereas he once only had one car, an old one, now he has many new ones. The one in which he takes the most delight is, significantly, a Corona, an artifact he treats as if it were literally as well as figuratively his crown. He entertains himself with meditations on the long list of options it exhibits. Throughout his life, it is the automobile that offers him the anonymous privacy in which to play at being himself, to view his world as though it were cinema. "As he sits snug in his sealed and well-assembled car the venerable city of Brewer unrolls like a silent sideways movie past his closed windows."[15] He recalls that as a child in the back seat of his father's Model A on the way to the New Jersey beach he "would look at the little girls on the sidewalks they drove alongside wondering which of them he would marry."[16] In his automobile Rabbit sings a song of himself. He is a mobile Leopold Bloom.

Rabbit uses other technological quotidiana to personate himself. He buys thirty Krugerrands with which he decorates the naked body of his wife Janice as part of their love play. It is obvious to us, and to him, that he takes more interest in the gold coins than in her body. He examines the medicine chest of his friends, the Webb and Cindy Murketts, speculating about their lives and his on the basis of the ointments, pills, and hardware he finds there. An earlier Rabbit, during his employment as a linotype operator for the local newspaper, the *Brewer VAT*, imagines accounts about the most intimate aspects of his life, set, in Updike's text and in Rabbit's fantasy, in the old-fashioned news-type that Rabbit works eight hours a day to produce.

Another of Updike's characters, Henry Bech, a blocked novelist seeking to avoid the consequences of his own intellectual sterility and mediocrity, also finds ways of personating himself by means of the articles and implements of his environment. He thinks of the single-handled hot and cold water faucet as a joystick on a biplane.[17] To break the dreariness of his life he accepts, uncharacteristically, an invitation to give a lecture at a girl's college in Virginia. Once there, he entertains himself by imagining the bed in his guest room to be a piece of pop art. "His bed, with its two plumped pillows one on top of the other like a Pop Art sandwich, its brocaded coverlet turned down along one corner like an Open Here tab on a cereal box, seemed artificially crisp and clean."[18] It is as if, by crawling between its covers, he can himself become a work of art. Suffering from writer's block now for six years, Bech designs a series of "repellent" rubber stamps. "HENRY BECH REGRETS THAT HE DOES NOT SPEAK IN PUBLIC. HENRY BECH IS TOO OLD AND ILL AND DOUBTFUL TO SUBMIT TO QUESTIONNAIRES AND INTERVIEWS." And finally, "IT'S YOUR PH.D. THESIS; PLEASE WRITE IT YOURSELF."[19]

2. *Authentication*: A second mode of self-stabilization by means of technological quotidiana is what I shall label "authentication." If personating is play, then authentication is struggle to identify, to prove, to test. Technological artifacts authenticate when they help us define ourselves and tell us who we are. Authentication is similar to personation, but a much more serious undertaking. Often it involves the generation of metaphors of self-measurement and personal change to render the inarticulate articulate and the unmanageable manageable. It, unlike personation, is goal-oriented.

Making love to his wife Janice, Rabbit seeks out a technological metaphor to authenticate his vigor: "He hasn't come with such a thump in months. So who says he's running out of gas?"[20] He sees his need for personal power and security reflected in changing automotive design. He thinks it an advantage that the new Japanese cars have lots of locks, especially on the gas tank.[21] He is apprehensive because he has heard that the auto parts stores have sold out of their gasoline siphons, a hardly recondite metaphor of his fear of having his own energies siphoned off by unknown individuals and organizations. He uses automobile metaphors to assess the ways in which he has aged. "When Rabbit first began to drive the road was full of old fogeys going too slow and now it seems nothing but kids in a hell of a hurry, pushing. Let 'em by, is his motto. Maybe they'll kill themselves on a telephone pole in the next mile. He hopes so."[22]

Bech, too, uses technological metaphors to measure and authenticate himself. Over a whiskey sour Bech tries "for one more degrading time to dig into the rubbish of his 'career' and come up with the lost wristwatch of truth."[23] He characterizes himself as an "uneasy, blurred composite, like the image left on film by too slow an exposure."[24] But even his depressing definition of himself in terms of a botched photographic image seems better than none at all.

The stabilization of authentication may also emerge from the recalculation of personal worth that follows disappointment. Delighted that an Anglican priest from Toronto has prepared a concordance of his work, Bech hopefully comments on the time and care that must have gone into the project: "It must have taken him, the priest, an immense amount of labor to compile such a concordance, even [Bech says with a proper measure of self-deprecation] of an *oeuvre* so slim. Ah, not really was the answer: the texts had been readied by the seminarians in his seminar in post-Christian kerygmatics, and the collation and printout had been achieved by a scanning computer in twelve minutes flat."[25] In this case, even though he had not wished it to be so, Bech's life is once again plotted according to the metes and bounds of technological artifacts. The lesson is not lost on him.

The automobiles in Rabbit's life serve as a point of the personal struggle between him and his son Nelson, who comes back home after three years of college. Nelson, who appreciates the aesthetics of television car smash-ups, manages to damage both his father's Corona and his mother's Mustang convertible. He has a deft understanding of the function of the automobile as talisman for his father, and he uses it as the focus of his oedipal struggle. He openly accuses his father of treating his automobiles as if they were magical.[26] And so they are. Nelson is of course the ultimate winner of the father-son struggle. He takes his father's Corona back to college at Kent State.

In the 1960s Rabbit is forced to come to terms with changes in technology that alter his most basic image of himself. He finds himself unemployed when his function as linotype operator is taken over by a computer. He makes mental notes of other, more subtle transitions in his technological environment: the "church key" is replaced by pull tabs and orange juice by a combination of chemicals called "Tang." Words fail him when he tries to describe an ancient advertising sign: is it an "un-neon" sign?

Bech uses technological means to explore the differences between him and the women in his life. He wonders if, far from distrusting machinery as they claim to, women are not in fact sexually excited by it. During an overnight train trip in the Soviet Union he makes the following journal entry:

Yerevan station at dawn. The women, puffy-eyed and mussed, claim night of total insomnia. Difficulty of women sleeping on trains, boats, where men are soothed. Distrust of machinery? Sexual stimulation, Claire saying she used to come just from sitting on vibrating subway seat, never the IRT, only the IND. Took at least five stops.[27]

Technological quotidiana may help us find out not only who we are, but where we are. Bech considers the differences betweens the electricity of his own culture and that of the Socialist countries he visits on a lecture tour.

Electricity was somewhat enchanted in the Socialist world. Lights flickered off untouched and radios turned themselves on. Telephones rang in the dead of the night and breathed wordlessly in his ear. Six weeks ago, flying from New York City, Bech had expected

Moscow to be a blazing counterpart and instead saw, through the plane window, a skein of hoarded lights no brighter, on that vast black plain, than a girl's body in a dark room.[28]

During an evening in a Rumanian nightclub, Bech is stunned by the size of the women's wristwatches, "which were man-sized, as in Russia."[29] He is particularly disturbed by the techniques of the driver who is assigned to him and his host, Petrescu, in Rumania.

When they went through a village, the driver would speed up and intensify the mutter of his honking; clusters of peasants and geese exploded in disbelief, and Bech felt as if gears, the gears that space and engage the mind, were clashing. As they ascended into the mountains, the driver demonstrated his technique with curves: he approached each like an enemy, accelerating, and at the last moment stepped on the brake as if crushing a snake underfoot.[30]

Complaining to Petrescu, Bech is asked if in the States he drives his own car. "'Of course, everybody does,' Bech said, and then worried that he had hurt the feelings of this Socialist, who must submit to the aristocratic discomfort of being driven."[31]

During a drunken walk in late-night Toronto he once again locates himself by means of quotidiana.

. . . block after block of substantial untroubled emptiness. He expected to be mugged, or at least approached. In his anesthetized state, he would have welcomed violence, but in those miles he met only blinking stop lights and impassive architecture. *And they call this a city*, Bech thought scornfully. *In New York, I would have been killed six times over and my carcass stripped of its hubcaps.*[32]

3. *Distraction*: A third mode of self-stabilization in Updike's novels is the distraction of the self from obligations, unpleasantnesses, and personal difficulties with the help of technological quotidiana. His characters buy time against these difficulties by musing, ruminating, and giving studious attention to what would otherwise be the most trivial and insignificant of technological details. They involve themselves in technological processes that would otherwise not have been parts of their lives. Rabbit, watching his house burn, fixates on the fact that the different materials of his home burn at different rates. Later, he distracts himself from that tragedy by noting that the only salvageable item is the television set. Bech, unable to write, does TV interviews. Bored with the table conversation at the Virginia girls' college, he finds himself reprieved by the clock. Pushed to the brink of a committment during a conversation about marriage with his lover, Beatrice, the phone rings.

"Without the telephone, Bech wondered, how would we ever avoid proposing marriage?"[33] Unable to avoid a trip to the Holy Land with his new wife Beatrice (quotidiana operating in this mode do not so much solve problems as buy the time to reconsider them), he distracts himself with ruminations about the overlay of commercial artifacts upon the ancient and sacred way of the Via Dolorosa: "Kodachrome where Christ stumbled, bottled Fanta where He thirsted."[34]

Rabbit, worried about the effects of the energy shortage on his livelihood, muses: "A new industry, gas pump shrouds."[35] But most of all, Rabbit distracts himself from the problems of relating to his wife and his son and the threatening aspects of what he takes to be big and unmanageable technology by means of a retreat into the technological minutiae of *Consumer Reports*. He uses that quintessential organ of technological trivia to avoid sex with his wife Janice. "He likes the upward glimpse of Janice's legs in the tennis dress as she goes upstairs" and makes a mental note that he should try making love to her some night when they are both awake. He could go upstairs at that very moment, but "is attracted instead by the exotic white face of the woman on the cover of the July *Consumer Reports*, that he brought downstairs this morning to read in the pleasant hour between when Ma went off to church and he and Janice went off to the club."[36] So instead of interacting with Janice, he reads about face cleansers, cottage cheese, air conditioners, compact stereos, and can openers.

Even when Janice attempts to initiate sex, Rabbit retreats into his minute world of technological quotidiania.

Naked, Janice bumps against the doorframe from their bathroom back into their bedroom. Naked, she lurches onto the bed where he is trying to read the July issue of *Consumer Reports* and thrusts her tongue into his mouth. He tastes Gallo, baloney, and toothpaste while his mind is still trying to sort out the virtues and failings of the great range of can openers put to the test over five close pages of print. The Sunbeam units were most successful at opening rectangular and dented cans and yet pierced coffee cans with such force that grains of coffee spewed out onto the counter. Elsewhere, slivers of metal were dangerously produced, magnets gripped so strongly that the contents of the cans tended to spatter, blades failed to reach deep lips, and one small plastic insert so quickly wore away that the model (Ekco C865K) was judged Not Acceptable. Amid these fine discriminations Janice's tongue like an eyeless eager eel intrudes and angers him.[37]

4. *Focus of Desire*: A fourth means of self-stabilization is the use of technological quotidiana to focus and define desire. At Vellum Press, the publisher of his novel *Think Big*, there is a young female assistant, "a quick black-haired girl fresh from Sarah Lawrence," who catches

Bech's attention. "Bech wondered if it was her hands that showed in the Xeroxes the firm sent him of his galley sheets."

Whoever it was had held each sheet flat on the face of the photocopier, and in the shadowy margins clear ghosts of female fingers showed, some so vivid a police department could have analyzed the fingerprints. Bech inspected these parts of disembodied hands with interest; they seemed smaller, slightly, than real hands, but then womanly smallness, capable of Belgian embroidery and Rumanian gymnastics, is one of the ways by which the grosser sex is captivated. He looked through the photocopied fingers for the hard little ghost of a wedding or engagement ring and found none; but then she might have been employing only her right hand.[38]

Later, Bech meets the young girl, whom he has theretofore only seen from afar or Xerographically mediated, at a lunch with his editor. As they shake hands in greeting, he holds on "a half-second longer than necessary to her hand. Her dear small busy clever hand. It was much whiter than in the Xeroxes, and decidedly pulsing in his."[39]

During Rabbit's clandestine examination of the Murketts' bedside table, he encounters "black-backed Polaroid instant photos."[40] Webb and Cindy Murkett, too, have used technological means to heighten and focus their desire.

The light in the room must have been dying that day for the flesh of both the Murketts appears golden and the furniture reflected in the mirror is dim in blue shadow as if underwater. This is the last picture; there were eight and a camera like this takes ten. *Consumer Reports* had a lot to say a while ago about the SX-70 Land Camera but never did explain what the SX stood for. Now Harry knows. His eyes burn.[41]

It is not only innocent desire at play that is focused by technological artifacts. The shame of guilty desire is defined and delineated as well. Deep in the dark years of the 60s, Rabbit, deserted by his wife, takes up with Jill, a young flower child who articulates anti-technological ideology by day and shares his bed by night. In this difficult and confusing situation Rabbit finds a quotidian artifact with which to measure the focal length of their relationship. Seeing in his own relation to Jill something of his country's shameful one to Vietnam, he identifies his semen with napalm, that artifact which most of us have never seen but which nevertheless became depressingly quotidian to those of us who watched American television between 1965 and 1972. Each morning Rabbit expects to find burns on her skin.

5. *Magic*: Finally, technological quotidiana function as implements of magic. There is a trivial and passive sense of magic in which any novel technological instrument, or any whose workings we do not fully

understand, any that appears to violate the common-sense science of the world as we have learned to perceive it, is called "magical." Electricity is still magical, in this sense, as is the Polaroid. It is not so much that technological men and women understand and therefore use these things; it is more accurate to say that they serve a function and therefore that they are used.

But there is a more profound sense of magic that is closer to the roots of technology in *techne*. There is "craft" in witchcraft. Tudor-Stuart England had numerous "cunning folk," practitioners of a kind of "white magic" that mixed pharmacologically effective remedies with verbal and other rituals. In a remark that provides an interesting contrast with Rabbit's assessment of his role as MagiPeel salesman, namely that it constituted a kind of innocent fraud, Monter suggests that their activities were "unofficial and even plainly illegal, but they were not necessarily fraudulent."[42] Margot Adler reports an interview with a contemporary witch named Cybele, to whom witchcraft was a form of technique: "If [we] have a law," she said, "that law would be: 'If it works, do it; if not, throw it out.'"[43] "I was shown how to do certain things, practical things. How do you make your garden grow? You talk to your plants. You enter into a mental rapport with them. How do you call fish to you? How do you place yourself in the right spot? How do you encourage them?"[44] Another witch tells her, "I learned from her that the Craft is a religion of hearth and fireside. The tools of the Craft are kitchen utensils in disguise. It's a religion of domesticity and the celebration of life."[45]

But magic is like other forms of *techne*; while it may work at the level of practice, its theories do not always adequately describe or explain what it does. This was a feature of *techne* noted by Plato on his visit to the artisans in order to find out if they were wise; and John Dewey remarked on it when he noted that the triumphs of the "scientific" revolution of Kepler and Galileo came as a result of ever more adequate practice with tools and artifacts which was wedded to sometimes obsolete theoretical apparatus, a metaphysics of contemplation of a finished universe.

In this sense, then, magic, like other forms of *techne*, involves the use of tools and artifacts according to a method in order to achieve certain desired results. What sets it apart from other forms of *techne* is its gnostic core, its recondite quality. Why is such practice recondite? It may be so because stronger and more established social institutions have attempted to foreclose on the forms such practice takes. The medieval church arrogated all magic to itself and condemned as heretics those who persisted in "freelancing." Similar situations have occurred in the history of the physical sciences and the medical arts. It may be so because special gifts exhibited by a talented minority are often resented and suppressed by a majority. Further, magic almost without exception constitutes attempts by individuals and subgroups to gain and stabilize power against systems and dominant groups, respectively. But perhaps the most significant feature of magic that sets it apart from other forms of *techne* is that both begin and end with the imaginative faculties, thus falling outside the realm of what has normally been taken to be the acceptably tangible in *techne*. Adler defines magic as "a convenient word for a whole collection of *techniques*, all of which involve the mind."[46]

Magic, as she found it practiced in the United States, "did not involve the supernatural. It involved an understanding of psychological and environmental processes; it was a kind of shamanism, a knowledge of how emotion and concentration can be directed naturally to effect changes in consciousness that affect the behavior [for example] of humans and [other animals]."[47]

A consideration of the results of the practice of magic, then, certainly brings us back to the central theme of this paper, the phenomenology of quotidian artifacts, particularly as means to self-stabilization between the extremes of subjectivity on one side and the overarching systems of nature and big technology on the other. Magic, like personation, authentication, distraction, and focusing of desire, involves the consolidation of power to the self in order to stabilize and protect it against decomposition.

But "magic," like "technology," is broadly and often contradictorily connotative. Some practices which claim to be magical do not in fact work. It is easier to make fake or bogus claims in the realm of the intangible than where immediate, visible, and well understood results are expected. Other practices which involve the claim of magic must ultimately yield to the effects of stronger magic. There is something of magic in the words (and in the use of the symbol-bearing artifacts) of the priest who unites two people in marriage. He has constructed a new entity, undertaken a social action. But there is also something of magic in the activities of friends and relations who may not wish to see the union succeed. Often their magic is stronger, and has its own store of symbol-laden quotidiana.

Updike has addressed the subject of technological magic in *The*

Witches of Eastwick.[48] His title includes a play on the word "witch" or "wicca," which many have suggested is from the Old English "wic," meaning to bend or turn. Thus, as Adler puts it, "a Witch would be a woman (or man) skilled in the craft of shaping, bending, and changing reality."[49]

Behind the somewhat fanciful facade of Updike's book there lies a serious and thought-provoking account of the nature of power and the struggle to obtain it. Each of the three witches is divorced, having transformed her husband into an inert object. Alexandra's husband had become less viable each day, shrinking "to the dimensions and dryness of a doll, lying beside her in her great wide receptive bed at night like a painted log picked up at a roadside stand, or a stuffed baby alligator."[50] "By the time of their actual divorce her former lord and master had become mere dirt – matter in the wrong place, as her mother had briskly defined it long ago – some polychrome dust she swept up and kept in a jar as a souvenir."[51] Jane's ex-husband hung in the cellar of her ranch house, among the dried herbs, and was occasionally sprinkled into one or another potion. Sukie had "permanized hers in plastic and used him as a place mat."[52] But during the course of Updike's fanciful account their power is taken away by a proponent of scientific technology, Darryl Van Horne, a mysterious scientist at work on a new technology for photovoltaic cells and an incessant collector of technological quotidiana which range from the apparatus of his lab to the plumbing of his hot tub to his meta-quotidiana, the pop-art objects that fill his house.

Before they meet Van Horne, the witches have successfully utilized domestic quotidiana to structure their respective selves and effect the accrual to them of personal power. But their magic ultimately falls prey to a stronger one, the magic of scientific technology and its practitioner, Van Horne. Their powers are as much diminished by it and by him as they are appropriated and set to serve purposes alien to the aims of the witches themselves.

Another of Updike's studies of magic is *The Coup*,[53] perhaps the most visibly "technological" of his accounts. Its central character is Hakim Felix Ellullou, the president of the African republic of Kush. The ironic and well-read Updike has given his protagonist a name that bears a striking resemblance to that of Jacques Ellul. Like the witches of Eastwick, Ellullou too struggles against Big Technology, in this case personified by the encroaching Americans with their mindless optimism and quotidian gadgetry. Ellullou fails to succeed in his task of self-stabilization precisely because he eschews the gadgets and methods that they bring, attempting to defend himself and his country against them by another kind of magic which combines Third World Marxism with Islamic fundamentalism.

For his part, Harry Angstrom had remained afloat in the sea of Big Technology by holding tight to the flotsam and jetsam of its quotidiana. It was both source and focus of his magic. But his is a solution particular to industrialized men and women. For Ellullou, Big Technology comes like a flash-flood in the desiccated wadis. The magical use of technological quotidiana which stabilizes Rabbit against its currents is unavailable to Ellullou because of his ideological commitments, which amount to a less powerful kind of magic. Ellullou attempts to exorcise the magic of Big Technology with theory, that is, with supernatural incantations, and fails. Rabbit deals with Big Technology, as with all else in his life, at the level of constantly readjusted practice. He at least stays afloat.

This has been a discussion of five modes of stabilizing the self against the forces that threaten its decomposition. For Arendt those forces include human subjectivity and the impersonal cycles of nature. To these I have suggested the addition of a third, the "second" nature I have called Big Technology. I have argued that beside the important studies of sociologists and anthropologists into the ways in which technological quotidiana serve this function, there lies in mainstream fiction a rich source of insights into the ways in which such activities come to fruition.

Texas A & M University.

NOTES

[1] Hannah Arendt, *The Human Condition* (Chicago, 1958).
[2] Arendt, p. 137.
[3] Arendt, p. 173.
[4] See, e.g., Jacques Ellul, *The Technological Society* (New York, 1964).
[5] Marshall McLuhan, *Understanding Media* (New York, 1964).
[6] Edward Hall, *The Hidden Dimension* (New York, 1969).
[7] Mihaly Csikszentmihalyi and Eugene Rochberg-Halton, *The Meaning of Things* (Cambridge, 1981).
[8] Joseph F. Coates, 'Computers and Business – A Case of Ethical Overload', in Larry Hickman, ed., *Philosophy, Technology and Human Affairs* (College Station, Texas, 1985), p. 383.
[9] Douglas Browning, 'Some Meanings of Automobiles', in Larry Hickman, ed., *Philosophy, Technology and Human Affairs* (College Station, Texas, 1985), p. 65.
[10] Browning, p. 66.
[11] John Updike, *Rabbit Run* (New York, 1960).

[12] John Updike, *Rabbit Redux* (New York, 1971).

[13] John Updike, *Rabbit Is Rich* (New York, 1981).

[14] Updike (1960), p. 93.

[15] Updike (1981), p. 28.

[16] Updike (1981), p. 126.

[17] John Updike, *Bech: A Book* (New York, 1970), p. 120.

[18] Updike (1970), p. 116.

[19] Updike (1970), p. 134.

[20] Updike (1980), p. 50.

[21] Updike (1981), p. 19.

[22] Updike (1981), p. 30.

[23] Updike (1970), pp. 141–142.

[24] Updike, (1970), p. 134.

[25] John Updike, *Bech Is Back* (New York, 1982), p. 53.

[26] Updike, (1981), p. 77.

[27] Updike (1970), p. 196.

[28] Updike (1982), p. 51.

[29] Updike (1982), p. 43.

[30] Updike (1982), p. 31.

[31] Updike (1982), p. 32.

[32] Updike (1982), p. 59.

[33] Updike (1970), p. 104.

[34] Updike (1982), p. 63.

[35] Updike (1981), p. 17.

[36] Updike (1981), p. 74.

[37] Updike (1981), pp. 45–46.

[38] Updike (1981), p. 141.

[39] Updike (1981), p. 143.

[40] Updike (1981), p. 284.

[41] Updike (1981), p. 286.

[42] E. William Monter, *Witchcraft in France and Switzerland* (Ithaca, New York, 1976), p. 173.

[43] Margot Adler, *Drawing Down the Moon* (New York, 1979), p. 73.

[44] Adler, p. 74.

[45] Adler, p. 73.

[46] Adler, p. 8 (my italics).

[47] Adler, p. 9.

[48] John Updike, *The Witches of Eastwick* (New York, 1984).

[49] Adler, p. 11.

[50] Updike (1984), p. 7.

[51] Updike (1984), p. 7.

[52] Updike (1984), p. 7.

[53] John Updike, *The Coup* (New York, 1978).

PAUL LEVINSON

SYMPOSIUM ON INFORMATION TECHNOLOGIES:

I. Impact of Personal Information Technologies on American Education, Interpersonal Relations, and Business, 1985–2010

Human beings are in the business of information. As Arthur C. Clarke pointed out,[1] we can last longer without food and water – longer as functioning individuals – than we can devoid of sensory connection to the outside world. Our bodies hum with information conveyed through a myriad of synapses, and these in turn get their cues from the wider nets we cast across ourselves, our communities, our planet, and beyond.

All organisms of course have means of perceiving the external world. But we humans, unsatisfied with the limitations of our naked senses, seeking to communicate beyond the confines of sight, speech, and hearing, invent technologies to extend our information almost as broadly and as swiftly as the mind itself can go.

The consequences of even the most seemingly modest of these information technologies in the past have been enormous. The change from picture-based hieroglyphic to phonetic alphabetic systems of writing (i.e., systems of writing which use a handful of arbitrary characters to record the sounds of speech rather than the images of the world), for example, has been linked to the rise of monotheism and the advent of science, mathematics, and logic in the Ancient World.[2] The mass dissemination of these letters nearly two thousand years later by the printing press in Europe helped set in motion the Protestant Reformation, the Scientific Revolution, the rise of national states, the development of capitalism, and a series of related upheavals in social and political life that brought us the Modern World.[3] Moreover, events of such magnitude are inevitably accompanied by ripple effects that permeate all aspects of human society. Thus, the avalanche of religious, scientific, and political tracts created by the printing press stimulated a powerful need to learn how to read, which in turn encouraged the rise of public education. A few historians have further related these events to the sharp dichotomy between childhood and adulthood that has characterized all but the most recent of times since the Renaissance.[4]

No one alive today can be surprised to learn that we are in the midst of an information upheaval of similar profundity, one that has been

177

Paul T. Durbin (ed.), Technology and Contemporary Life, 177–191.
© 1988 *by D. Reidel Publishing Company.*

going on since the invention of telegraphy and photography some hundred-and-fifty years ago. The most recent component in this electro-chemical revolution is the computer – or, more specifically, the capability of instant, interactive, global communication of text in the privacy of one's home made possible by microcomputers, modems, and telephone networks.

This paper will examine some of the likely consequences of personal or micro computing and telecommunication for instructional and commercial activities in America and the world at large in the next thirty years. The analysis will extend to secondary impacts of this technology on the family and modes of urban living, and will refer to related technologies such as telephone, television, and video conferencing.

PERSONAL COMPUTER CHARACTERISTICS

Computers in one form or another have performed at the cutting edges of science, engineering, and business since the late 1940s, but their penetration of small offices and individual homes awaited the development of microchip technology (specifically the Z80 8-bit chip) in the mid 1970s. The impact of communications media in the home has often been underestimated in history, with the telephone mistaken in the first few decades of its existence as just a telegraph with voices, and TV as just a smaller, blurrier version of the motion picture screen. In both cases, such misperceptions cost the established bases of media power dearly. Thus, Western Union Telegraph's view that the telephone was little more than a "toy" led it to sit idly by until the telephone far outstripped the telegraph in importance, while the underestimation of television by Hollywood moguls all but put their sort of movie industry out of business.[5]

The power of communications media in the home lies in their ability to bring public events and activities into the privacy and comfort of our living quarters.[6] Telephone internalizes the intrinsically external activity of conversation, and radio and TV do the same for news. Personal computers (equipped with modems hooked into telephone lines) do for the written word what telephone has done for the spoken word and TV has done for images.

This capacity to engage the external world from our armchair, living room, or bedroom increases both our authority or ability to manipulate external events, and our vulnerability or likelihood to be the object of external manipulation. Thus, the price we pay for the convenience of speaking to a far-away friend or important business associate by phone is the unwanted call from a nuisance relative or salesperson, and television exposes us to commercials as well as news.

The question concerning the impact of personal computers in the next twenty-five years is to what degree they will simply extend the pros and cons of media in the home as represented by telephone and television, and to what degree they will break new ground in our various relationships with the external world. Or we can put the question as follows: will the ability to move written text in and out of the home, instantly and globally *via* computers, result in social configurations substantially different from those engendered by prior electronic media? The balance of this analysis will explain and detail the ways in which the answer to this question is likely to be yes.

ELECTRONIC FACILITATION OF PRINT

Electronic media are commonly looked upon as competitors or successors of the printed word and all that a print-based culture entails.[7] While this view recognizes the undeniable way that sounds and images tend to upstage the products of print, it also overlooks an equally profound tradition of electricity and electronic media in the service of print culture. The electronic transmission of printed characters by the telegraph, for example, utterly revolutionized and stimulated the new profession of journalism in the middle of the last century, and the introduction of the electric light, in the opinion of at least one historian, did more to increase the habit of reading than any other invention since the press itself.[8]

The creation and transmission of printed characters by computers is very much in the second tradition. Freed from its formerly inextricable marriage to the paper page, print can now sail through the world with the swiftness and ease of a voice on the phone or an image on TV. Moreover, unlike the voice on the phone, print received on the computer screen does not require an immediate response, and unlike the TV images which we can only receive and not create, the print that moves through computer hook-ups is easily created as well as received by all parties to the communication. As we will see below, these differences between computer and traditional modes of telecommunication are especially significant for education.

This new revolution in print entails much more than mere word-processing. Although lightning manipulation of printed characters on a screen greatly aids the process of composition, once the results of that process are committed to a paper page – *via* printer connected to the computer or whatever – that page enters our culture in much the same way as any traditional product of the press. Thus, for anyone reading this report on xerox pages, the fact that I composed this on a Kaypro II computer would be of little consequence. On the other hand, were that person and I members of the same computer network or otherwise linked *via* computer at the time I wrote this report, it could be read in the comfort of someone's home a mere few minutes after the report had been completed. (Indeed, for those interested in what the electronic as opposed to the paper transmission of print feels like, I have a copy of this report stored on the Electronic Information Exchange System operating out of the New Jersey Institute of Technology in Newark, N.J. My number there is 1303 – anyone can message me and I will be glad to send a copy of this report electronically.)

The combination, then, of word processing and telecommunications provides a pipeline from intellect to intellect that is unparalleled in its immediacy and intimacy. Late at night, when ideas run rampant but the body is ordinarily too fatigued for any sort of intellectual exchange, fingers too tired to write a letter or even hold a phone flutter lightly over keyboards, committing to the screen the deepest contents of the mind. With another flick of the finger such ideas are sent speeding across continents, to be received and pondered at a time of the receiver's convenience. The implications of such electronic transfusions are immense for education, business, and family life.

IMPACT ON EDUCATION

Since the inception of education with the ancient private tutor, through its development from the one-room school house to the teeming classrooms of today's huge universities, the essence of education has been words: words spoken by teachers and students, words printed in books, words written by students, words of all sorts. A picture is worth a thousand words only with some person or caption to explain it – only when framed in words.

But the classroom, for all its charm and venerability, is not the most efficient medium for the exchange of words. To begin with, people expected to communicate at appointed times and places may not always be willing or able to communicate effectively at those times. A teacher may have an off day; students may be distracted for any number of reasons. Further, much of what does get communicated in classrooms is lost moments after its communication. Even the fastest note-taker misses some of the give-and-take of lecture and discussion (not to mention being locked out of the discussion at the time the notes are being recorded); and even the most conscientious instructor inevitably overlooks or shortchanges a student question or two. Tape recorders may remedy some of this, but they also can be an intrusive presence in the classroom, on occasion inhibiting the free flow of discussion.

Words exchanged through computers suffer from none of this. They are written and read at times convenient to the writers and receivers. In a typical computer "classroom," faculty compose lectures, comments, and questions at their home terminals, and then transmit these *via* telephone wires to a central computer. There they are stored for students, who access the faculty material by calling up the central computer from their own home terminals. Telecommunications software permits students to "download" or record the faculty material for further or later reading, after which students can enter questions and comments of their own in the central "class." This "asynchronous" procedure – reading and responding at your own pace and time – seems to heighten both the quality and frequency of student participation.[9] Moreover, the central computer provides a permanent record of all proceedings, a safety net for retention of the off-hand but valuable aside that might easily get lost in the shuffle of voices in the live classroom.

The above is not a future scenario: education *via* computer conferencing is being conducted right at this moment on more than a dozen computer networks.[10] Of course, before the computer can supplement conventional education in any widespread way – let alone replace it – students must be equipped with computers and modems and know how to operate them. Word processing and telecommunication by computer must become as second nature to us as scribbling with a pencil or talking on the phone.

Universities are beginning to break the ice on the upper end of the educational spectrum; several now provide a computer system to every enrolling student.[11] But the real breakthrough will come when students are computer-knowledgeable *before* they enter the educational process – when computers permeate our homes and societies to the point that the average child is as comfortable with a keyboard and screen as with crayons, pencils, and televisions. Recent figures put the number of personal computers in use in the U.S. at approximately nineteen million; the number is expected to quadruple by the end of the decade. (About twenty percent of these use modems. This percentage can be expected to increase radically as computer manufacturers include modems as part of the internal hardware of machines they sell.)[12]

When considering the numbers of personal computers in use and their applicability to education via computer conferencing, however, we must be careful to distinguish the under-$500 computer fit primarily for games and recipe lists (e.g., the Commodore 64) from computers that currently sell from around $1000 (e.g., IBM PCjr, Apple IIe, Kaypro II, etc.) and are capable of sophisticated word processing and telecommunications. The first type of computer is often purchased in a fad-like way, and frequently winds up collecting dust in the attic as the family tires of the electronic games. The second is the entrée to the serious and sustained use of computers in schoolwork and business, and often leads to the purchase or use of even more sophisticated equipment.[13]

As suggested above, computer teleconferencing can be used to teach any subject whose content can be conveyed in words. This of course would include traditional liberal arts and primary and secondary school areas such as English, History, Philosophy, Social Studies. Sciences requiring hands-on laboratory work and art and music appreciation would present special problems, although the latter could make use of video conferencing to supplement the computer. In addition to telecommunications, computers can assist the student *via* software programs that check spelling, drill in history, etc. Indeed, the development of CAI ("computer assisted instruction") software is one of the fastest growing areas in education.[14]

Like all evolutionary developments – biological as well as technological – the ascension of a new species of organism or machine is never an unmitigated blessing.[15] Despite the drawbacks of the in-person classroom, it has worked reasonably well for hundreds of years, and has advantages (e.g., availability of nonverbal cues) that will be lost in the symbol-only computer environment. One remedy is to employ computer conferencing not as a substitute but as a supplement to in-person education (or perhaps make the live classroom supplement the computer). Another response is to ask of computer conferencing and computer assisted education, not whether it is an improvement upon the live classroom in every way, but whether it represents a net gain of some sort in educational power. Educators and those concerned with learning will no doubt be debating these questions for some years to come. My guess would be that by 2010 the issue will have been firmly decided on the side of computers.

IMPACT ON INTERPERSONAL RELATIONSHIPS AND FAMILY

One apparent shortcoming of education and indeed any activity conducted *via* computer teleconferencing is the lack of person-to-person social interaction that ordinarily would accompany such activities. At the lower levels of the educational system, the socialization of children – their capacity to make and sustain friendships, work in groups, etc. – is at least as important a product of education as the ability to read and write and add and subtract. Further up the scale, the romance on a campus, the late night talks in coffee houses and pubs, are an essential part of university life.

Computer relationships, however, do not so much eliminate these zests of life as situate them in a different but nonetheless supportive medium. As is the case with education itself, the essence of social interaction is exchange of information through words. Friendships, romances, group dynamics of all sorts run on words, and these can be communicated almost as easily by computers as by mouths. (In some respects more easily by computers, as they remove the communicator from some of the anxieties of in-person communication.)

Again, the development of personal relationships through computers is more than mere prediction: a recent survey of user activity on "Parti on the Source" (one of the largest commercial computer networks in the country), for example, reported that a discussion of "Electronic Love" had more entries (nearly three thousand in less than a year) than any of the other hundreds of group discussions conducted on the Parti system.[16] Computer teleconferencers are apparently not only creating strong and deep friendships on-line, but are falling in love, and in some cases even engaging in quite explicit verbal sexual activities *via* modem and computer.[17] (Of course, in the latter cases, some in-person physical supplement might be more necessary than in predominantly educational uses.) None of this should really come as any surprise, or be regarded as unnatural or deviant. People, after all, have been falling in love through hand written letters for centuries, and more recently the telephone has played an important role in teenage romance and sexual activity.[18]

Human relationships move on information, of which direct physical contact is but a contributing part.

Pursuance of interpersonal relationships *via* computer may be beneficial to traditional family life. What parent would prefer seeing his or her child initiate romances and relationships at all night parties and singles bars when such initiations could occur *via* computers in the safety of one's home? The decentralization of social interaction made possible by computers continues the decentralization of entertainment begun by older electronic media such as radio and television, which despite the frequent criticism heaped on them have done much to keep members of the family at home.[19]

Telephone, another irritant to many parents and social observers, has similarly situated many social activities in the home.

Of course, the faceless quality of computer communication not only lacks a significant human dimension, but is vulnerable to unique sorts of abuse when used for personal relationships. An oft-told story in computer conferencing circles, probably not apocryphal, tells of a middle-aged Los Angeles businessman who fell deeply in love with a woman he communicated with in New Jersey. So in love was the gentleman that he was about to leave his wife and three children, and take the next plane back East. Only repeated pleas by the parents of the New Jersey "woman" stopped him: she was fourteen years old and handicapped. Words devoid of face and even voice can certainly be used to deceive.

As is the case with education, those who wish to use the computer for interpersonal relationships will have to decide whether the pros outweigh the cons, whether the capacity to communicate with virtually anyone on the face of the Earth, instantly, through the non-demanding and usually dignified medium of written words is worth the potential for deception inherent in all word-only communication.

IMPACT ON BUSINESS AND URBAN LIFE

The use of word processing, data management, and telecommunication capabilities of personal computers is far better established in business than in education and interpersonal relations, and thus already has been the subject of several scholarly and popular works.[20] The key ingredient, as in education and social relations, is the informational nature of most business and commercial activity: if the essence of a job is the transfer of information, as it usually is, and the transferring of information is less expensive than the transporting of physical bodies in order to transfer the information, then the telecommunicating office makes economic sense. The result is the propagation of electronic cottages and work-at-home offices.

This decentralization of office work – dispersion would probably be a better term – will have significant impact on the social structures of offices and family life. Removal of employees from the physical proximity of supervisors will undermine traditional social controls on workers – already weakened by "flexitime" procedures – and indeed may eliminate the need for certain types of middle-level supervisory management (e.g., office managers whose job is to make sure that workers come in on time). Reduction of social control, however, may well result in an increase in worker productivity, as workers devote less energy to pleasing supervisors in a social sense, and more toward the literal output of words which will be the only measure of employee performance in the computer-dispersed office.

The work-at-home office will also have an enormous effect on family relations. Unlike the traditional one-worker family, in which the husband goes off to work all day and leaves the wife and kids at home, or the more recent two-worker family, in which husband and wife go off to work at their separate jobs, the computer-working family will be in close physical proximity all day. The parallel location of children in the home *via* computer-conducted education and parents in the home *via* computer-dispersed offices is likely to result in an era of family togetherness unknown since the days of the family living and working together on the farm. (Nor is such a return to past sharing likely to be universally acclaimed: at least one survey reports an increase in matrimonial disputes in families that work at home *via* computer.[21])

The movement of information rather than bodies to work may have drastic consequences for general city life as well. Computers that keep people at home will at first seem an unmitigated boon to urban life, as pressure is reduced on the overcrowded roadways and crumbling mass transit systems. Beyond a limited percentage, however, the reduction of the physical work force will jeopardize the heart of every city's power: real estate. Who will work in the offices of city skyscrapers with more and more people electing to work at home from computers? Who will pay to rent these offices as company after company shrinks its physical office space in accommodation of the new electronic business environment?

Offices may well be converted into housing units, provided that enough people still have an interest in living in the city. Indeed, with most forms of entertainment available electronically, about the only service unique to the big city would be restaurants: food, after all, is one type of energy that cannot be broadcast or transmitted through cables. The computer thus may transform our urban centers into cities of restaurants, with a few department stores and shopping malls thrown in for color. (Physical manufacturing plants also deal in tangible, non-informational commodities, but high real estate costs forced most such plants to leave the urban centers years ago.)

Is there anything that we as individuals or as a society can do to stop this process or slow it down, assuming we would want to? People like to think so, but the long stretch of recorded history discloses virtually no cases in which a society has rejected a technology as far advanced as the computer now is in ours.[22] On the other hand, if improvements in the quality and duration of human life are any indication, we would do well to bear in mind that the consequences of such technological inevitability have on the whole been quite beneficial to our species.[23]

SUMMARY AND CONCLUSIONS

The crux of the revolution in personal computers is the potency and ease with which they allow us to create, manipulate, transmit, and receive written words, and thus this revolution transforms all affairs of the intellect and many of the heart. If print was the "gunpowder of the mind,"[24] then the boost that computer word processing and telecommunication give to intellectual activity is something akin to nuclear power.

The most fundamental and far-reaching consequences of this intellectual empowerment will be in the field of education, long an arena for the exchange of spoken and written words. In-person classrooms, obliging teacher and students to conduct these exchanges at rigidly set times and places, will first be supplemented and then probably replaced entirely by computer conferences that allow teacher and students to exchange words and information from the comfort of their homes. Such electronic classrooms are already beginning to operate on the higher levels of education, and will probably work their way down to the primary levels by the middle of the next century. In-person classes will eventually have much the same relationship to computer conference classes as Broadway plays and live musical performances now have to film, TV, and recordings: the live events will be very special occasions, rare occurrences for specific purposes rather than common modes of activity.

Much the same will occur in the world of business, where personal computers are already well established. Most businesses operate on information, and the inexpensiveness of information movement as opposed to people movement will exert continued pressure toward keeping people at home and sending their information *via* computer.

These changes in education and business, reinforced by video conferencing and other electronic technologies in cases where live images are needed, will result in a resurgence of familial proximity, as children and parents do an increasing amount of their life's activities in the home. Indeed, the swift and global conveyance of information from homes *via* computers will likely revive a conception of family lost since the Industrial Revolution.

The image of people interacting with the world from the convenience of their homes need not be an image of people locked in their homes, or in any sense prevented from experiencing the external world and its pleasures in person. New media and technology offer options, and although these options are usually difficult to resist, they rarely preclude the use of earlier media or totally obliterate prior patterns of living. People, after all, elect to send letters even in this age of electronic communication, and those who dislike flying can always take a boat, train, or drive to their destination. The result of the computer revolution in telecommunications will be not that people are confined to their homes, but that they will only leave their homes if and when they want to.

Fairleigh Dickinson University and the New School

NOTES

[1] 'Communications in the Second Century of the Telephone', in *The Telephone's First Century – and Beyond*, Preface and Afterword by John D. deButts, Introduction by Thomas E. Bolger (New York: Crowell, 1977), p. 86.

[2] See, for example, Harold A. Innis, *The Bias of Communication* (Toronto: University of Toronto Press, 1951), Marshall McLuhan, *The Gutenberg Galaxy* (New York: Mentor, 1962), and Eric Havelock, *Preface to Plato* (Cambridge, MA: Harvard University Press, 1963). See also Chapter 6 in my *Mind at Large: Knowing in the Technological Age* (Greenwich, CT: JAI Press, forthcoming 1987).

[3] See Innis, *op. cit.*; and Elizabeth Eisenstein, *The Printing Press as an Agent of Change* (Cambridge: Cambridge University Press, 1979). See also *Mind at Large, loc. cit.*

⁴ On the connection between print, education, and childhood, see Neil Postman, *Teaching as a Conserving Activity* (New York: Delacorte, 1979), p. 45, and Joshua Meyrowitz, *No Sense of Place* (New York: Oxford University Press, 1985), Chapter 13. See Meyrowitz also for a consideration of the impact of non-print electronic media (e.g., television) on current conceptions of childhood.

⁵ See S. H. Hogarth's 'Three Great Mistakes', *Blue Bell*, November 1926 for a colorful account of Western Union's folly; Robert Sklar's *Movie-Made America* (New York: Vintage, 1975), pp. 269–286, provides details on the Hollywood moguls' decline. See my 'Toy, Mirror, and Art: The Metamorphosis of Technological Culture', in Larry Hickman and Azizah al-Hibri, eds., *Technology and Human Affairs* (St. Louis: Mosby, 1981), pp. 56–65, for an analysis of the general tendency to regard new technologies as toys.

⁶ As Marshall McLuhan put it, "North Americans may be the only people in the world who go outside to be alone and inside to be social." See his 'Inside on the Outside, Or the Spaced-Out American', *Journal of Communication*, Autumn 1976, pp. 46–53.

⁷ This is due mainly to the work of McLuhan, *The Gutenberg Galaxy*, *op. cit.*, and *Understanding Media* (New York: Mentor, 1964), and others such as Lewis Mumford, *The Myth of the Machine*, vol. 2: *The Pentagon of Power* (New York: Harcourt Brace Jovanovich, 1970). The two agreed on the conflict between electronics and print, but disagreed on the outcome and whether that outcome would be beneficial to humanity.

⁸ David de Hahn, *Antique Household Gadgets and Appliances* (Woodbury, NY: Barron's, 1977), p. 121.

⁹ The average student participation rate of twenty percent in a traditional live class is sometimes more than trebled in a computer telecommunicated class, according to remarks made by Murray Turoff (Director of the Electronic Information Exchange System), Computers in English and the Humanities Conference, New Jersey Institute of Technology, Newark, N.J., February 23, 1985. For more on differences between in-person and computer mediated classes, see Starr Roxanne Hiltz, 'The Virtual Classroom', in M. Turoff and M. Heimerdinger, eds., *Communication, Computers, and Education* (Norwood, NJ: Ablex, forthcoming); Christopher Dede *et al.*, 'Communications Technologies and Education: The Coming Transformation', in Howard Didsbury, Jr., ed., *Communications and the Future* (Bethesda, MD: World Future Society, 1982), pp. 174–182; and Lawrence Welch, 'Using Electronic Mail as a Teaching Tool', *Communications of the ACM*, February 1982, pp. 105–108. A serious weakness in most of these studies is that the classes conducted *via* computer conferencing were not for formal academic credit (students were not graded, etc.), and thus are not fully comparable to the typical in-person classroom situation.

¹⁰ Courses are currently being offered on such national computer networks as the Electronic Information Exchange System, Parti on the Source, Parti at NYIT, Arpanet, ConferII, and numerous smaller local networks. For a general description of the role of computers in education, of which classes conducted by computer teleconferencing are but a part, see John Goodlad, *A Place Called School* (New York: McGraw-Hill, 1984); and the April 1984 issue of *Computer*.

¹¹ Examples are Drew University which provides Epsons, Stevens Institute of Technology which furnishes DEC-350s, and Drexel University which includes a Macintosh as part of every student's entrance package. Other universities offer computers – usually Apples – to students at substantial discounts, or set up personal computer workstations available to all students. For details, see Alfred Lee's guest editorial in *Infoworld*, October 8, 1984.

[12] The figures were obtained in a phone interview with The Future Computing Company in Dallas, Texas. Projections of future modem usage were not included. The growing number of personal computers that come equipped with modems (including models sold by such popular companies as Apple, Radio Shack, and Kaypro), however, seems ample evidence that the use of computers for telecommunication will continue to rise.

[13] The distinction between these two types of computers explains the seemingly schizophrenic view that "the bottom has fallen out" of the home computer market, and the microcomputer is selling better than ever: clearly the under-$500 models are suffering, and the $1000-and-over machines are thriving. Indeed, Alfred Lee suggests that we reserve the name "home" computer for the first type of micro, and the name "personal" computer for the second. See his 'Overview' paper presented at the Computers in English and the Humanities Conference (see above, note 9).

[14] See, for example, Eva Thury, 'Organizing By Computer', paper presented at Computers in English and the Humanities Conference for a discussion of how software programs on Macintosh computers are used to assist students in English and History courses at Drexel University.

[15] For more continuities and analogies between biological and technological evolution, see my 'Human Replay: A Theory of the Evolution of Media', Ph.D. dissertation, New York University, 1979 (Ann Arbor, MI: University Microfilms); 'Evolutionary Epistemology Without Limits', *Knowledge: Creation, Diffusion Utilization*, June 1982, pp. 465–502; 'What Technology Can Teach Philosophy', in my *In Pursuit of Truth* (Atlantic Highlands, NJ: Humanities, 1982), pp. 157–175; 'Information Technologies as Vehicles of Evolution', *Technology in Society*, vol. 6, 1984, pp. 193–206; 'Technology as the Cutting Edge of Cosmic Evolution' in Paul T. Durbin, ed., *Research in Philosophy & Technology*, vol. 8 (Greenwich, CT: JAI, 1985), pp. 161–176; and *Mind at Large*, *op. cit*. See the bibliographies in these publications for lists of other research in this area.

[16] The survey was conducted by Michael Gilsen, and appears in his 'Super Index' Conference, Answer 42, on Parti on the Source. The closest runner-up to 'Electronic Love' was 'Gay Rights', with just over two thousand entries.

[17] Unconfirmed comments signed with pseudonyms in the 'Electronic Love' and similar discussions on Parti, CompuServe, and other networks describe highly explicit verbal sexual exchanges sometimes accompanied by masturbation. See also Robert Francoeur's report on sexuality and computers presented to the Mellon Interdisciplinary Seminar on Effects of New Technology, Fairleigh Dickinson University, May 1984.

[18] See Man-Kong Lum, 'A Study of the Impact of the Telephone on Human Sexuality', M.A. thesis, The New School for Social Research, May 1984, and the extensive bibliography at the end of that document.

[19] See, for example, Paul Kurtz, 'Some People Believe Anything They See on TV', in George Rodman, ed., *Mass Media Issues* (Science Research Associates, 1984), pp. 216–219, for typical criticism of TV as destructive of traditional social structures; see also Postman, *op. cit.*, p. 209, for the view that television fosters "individualized" attitudes that undermine family and community. These critics fail to recognize that whatever the anti-social effects of TV programs, however personally engrossing and unsociable the actual process of TV viewing may be, this process usually takes place in the home and thereby works to support the structure of family in a very profound way.

[20] Alvin Toffler's *The Third Wave* (New York: Morrow, 1979) was one of the first to popularize the notion of working at home through computers. See Starr Roxanne Hiltz,

Online Communities: A Case Study of the Office of the Future (Norwood, NJ: Ablex, 1984), for a more scholarly approach.

[21] Hiltz, *loc. cit.* Also reported in 'The School of Management and Strategic Studies Report' (publication of the·Western Behavioral Sciences Institute, La Jolla, CA), Spring 1984.

[22] Indeed, about the only known case of any society's rejection of an available major technology was Samurai Japan's shunning of the gun, described in Noel Perrin's *Giving Up the Gun* (Boston: Godine, 1979). In the long run, of course, even this rejection proved to be temporary, and was followed by a period of rapid military industrialization. See William McNeill, *The Pursuit of Power* (Chicago: University of Chicago Press, 1982), pp. 112 and *passim*, for more on the medieval and modern military history of Japan. Poison gas seemed to be an example of an astutely rejected technology in our own time, until its recent re-use in the Iran–Iraq conflict. Technologies thus appear beyond our capacity to banish outright. But this does not mean that technologies are beyond our control. For a discussion of how we can control technologies through the deployment or countervalence of other technologies, see my 'Media Evolution and Rationality as Checks on Media Determinism', in Sari Thomas, ed., *Studies in Mass Communication and Technology*, vol. 1 (Norwood, NJ: Ablex, 1984), pp. 231–238. The invention of window shades to control the lack of privacy created by windows is an archetypical example.

[23] Scholars who see sustenance in technology have been far outnumbered by critics, but have been ably represented by writers such as R. Buckminster Fuller, *Utopia or Oblivion* (New York: Bantam, 1969); Samuel Florman, *The Existential Pleasures of Engineering* (New York: St. Martin's, 1976); Pierre Teilhard de Chardin, *The Future of Man* (New York: Harper & Row, 1959); and Colin Cherry, *The Age of Access* (London: Croom Helm, 1985). See my 'Human Replay', *op. cit.*, Chap. 9, and *Mind at Large, op. cit.*, Chap. 9, for speculation on the ultimately positive evolutionary significance of technology for humanity and the universe as a whole.

Twentieth-century criticism of technology has generally taken the forms of: (a) technology deprives us of our humanity in some fundamental spiritual sense (e.g., Karl Jaspers, *Man in the Modern Age*, trans. E. & C. Paul (London: Routledge & Kegan Paul, 1951; German original, 1931), pp. 44–45: in an advanced technological world such as ours, "man cannot remain man"; and Jacques Ellul, *The Technological Society*, trans. J. Wilkinson [New York: Knopf, 1964; French original, 1954], p. 321: the human being "was created for a living environment, but he dwells in a lunar world of stone, cement, asphalt"); and (b) technology destroys or seriously impairs our capacity for rational reflection and analysis (again Ellul; also Herbert Marcuse, *One-Dimensional Man* [Boston: Beacon, 1964], p. 18: "The techniques of industrialization . . . prejudge the possibilities of Reason"; and Lewis Mumford, *The Pentagon of Power* [New York: Harcourt Brace Jovanovich, 1970], p. 298: "to be aware of only immediate stimuli and immediate sensations [what Mumford takes to be the environment fostered by electronic media] is a medical indication of brain damage"; see also Marx Wartofsky, 'The Critique of Impure Reason II,' *Science, Technology, & Human Values* (Fall, 1980), pp. 5–23, for summaries and discussion of Habermas, Althusser, and others on the technological constriction of reason.)

I see the following problems with these arguments: (1) Since all technologies originate in human minds, the claim that technology renders us incapable of being ourselves is on the face of it wrong. Technology has transformed earlier human conditions to our present expressions of humanity, to be sure; and one might mount an argument that these earlier states are preferable to our present conditions; but such an argument compares one type of human condition to another, rather than an earlier human to a somehow non-human technological condition. (Further, if we take the preservation and extension of human life to be a good – which I do – then the current human condition is a vast improvement over prior conditions with average lifespans of thirty-five years, etc.) (2) The claim that technological societies impair or even restrict reason is undermined by this very criticism of technology – which, if not correct, is certainly the product of considerable rational analysis. Interesting that the carnage of the technological society somehow always stops short of prejudicing the mentality of those who pounce to criticize it. (3) Further evidence of the vitality of our rationality in a technological society is found in our capacity to develop new technologies which quite deliberately redress shortcomings of current and older devices. Thus, the ephemerality of electronic media lambasted by Mumford has since been rectified by electronic media such as video tape – and indeed, electronic transmission of text – which act to preserve the products of communication. Such remedial media, as I call them, are tangible expressions of our rational direction of technology. (See last two sentences in note 22, above.)

See *Mind at Large* for further discussion of the inadequacies of modern criticism of technology. Sadly, the assumption in many academic circles is that a pro-technological attitude is a naive, innocent position uninformed by such faulty critiques; in fact, supporters of technology such as Fuller, Florman, and myself operate from a post-pessimistic position. The easy sophistication and appeal of pessimism is understandable, but the serious study of technology could be greatly improved by more careful attention to some of its theoretical champions, and by citation of our works as rigorously as we are expected to cite the works of the critics.

[24] David Riesman, quoted in Postman, *op. cit.*, p. 65.

S. MUTHUCHIDAMBARAM

SYMPOSIUM ON INFORMATION TECHNOLOGIES:

II. Information Technology, Citizens' Rights, and Personnel Administration

INTRODUCTION

Article 12 of the Universal Declaration of Human Rights proclaims that: "No one shall be subjected to arbitrary interference with his privacy, family, home or correspondence, nor to attacks upon his honor and reputation. Every one has the right to the protection of the law against such interference or attack." Interference with or attack on such a privacy may originate from various institutions for different reasons. With the advent of the new technology, informatics or datamation has become an industry by itself, thereby accelerating the already existing hunger for more data about people and personnel. Supply creates demand. The (Canadian) Federal Task Force Report, *Privacy and Computers* (1972), has acknowledged that: "More personal information is being collected than most Canadians probably suspect, and is made available to a larger number of users than is probably supposed. There is no evidence that either of these trends will decline." Developments in information technology since that report have brought us to the threshold of Orwellian proportion of *Nineteen Eighty-Four*. There is an inverse relationship between the information demands of "big brother" (the state) and "small brother" (complex formal organizations) on the one hand and individual privacy, autonomy, and dignity on the other. For these reasons, privacy is a matter of social concern.

The purpose of this paper is to explore the various dimensions of computerized personal and personnel data and their policy implications. This is an exploratory study and should be treated as such.

PRIVACY: CONCEPT AND MEANING

Privacy is not a single value.[1] It is a multidimensional concept[2] or a constellation of values, claims, and interests in a universe of concurring and competing values, of supporting and antagonistic claims, of allied and adverse interests. It is time-bound and culture-bound. It has pragmatic and normative aspects. The consciousness with which claims of privacy

193

Paul T. Durbin (ed.), Technology and Contemporary Life, 193–215.
© 1988 by D. Reidel Publishing Company.

are made and the situations in which privacy is granted and practiced jointly determine the normative and pragmatic aspects of privacy. "Thus, situationality necessarily reflects the historical traditions of the society, the society's technology, the individual's ecological setting, the stage of the individual's life cycle, the individual's sense of self, the nature of the individual's inter-personal relationships."[3] There is also an historic dialectic between the views of privacy from the standpoint of the individual and from the standpoint of the state or other institutions.[4] It is a dialectic process involving differential accessibility or a balancing of opening and closing of the self to others over time and with changing circumstances.[5]

On a comparative basis, privacy has an element of both universality and particularism. Privacy as a generic concept occurs in all cultures but it also differs among cultures in terms of the behavioral mechanism used to regulate desired levels of privacy.[6] In tracing the universal and biological basis of privacy, Alan Westin has made the following observation:[7]

Man likes to think that his desire for privacy is distinctively human, a function of his unique ethical, intellectual, and artistic needs. Yet studies of animal behavior and social organization suggest that man's need for privacy may well be rooted in his animal origins, and that men and animals share several basic mechanisms for claiming privacy among their own fellows. . . .

One basic finding of animal studies is that virtually all animals seek periods of individual seclusion or small-group intimacy. This is usually described as the tendency toward territoriality, in which an organism lays private claim to an area of land, water, or air and defends it against intrusion by members of its own species. A meadow pipit chases fellow pipits away from a private space of six feet around him. Except during nesting time, there is only one robin on a bush or branch. . . . For species in which the female cannot raise the young unaided, nature has created the "pair bond," linking temporarily or permanently a male and a female who demand private territory for the unit during breeding time. Studies of territoriality have even shattered the romantic notion that when robins sing or monkeys shriek it is solely for the "animal joy of life." Actually, it is often a defiant cry for privacy, given within the borders of the animals' private territory to warn off possible intruders.

Secrecy is another concept which appears to be similar to privacy since both involve boundaries and the denial of access to others. But they differ in the moral content of the behavior which is concealed.[8] Privacy is consensual where secrecy is not necessarily so. There is a "right to privacy" but there is no established or agreed-upon "right to secrecy."[9] Secrecy implies the concealment of something which, in general, is negatively valued by the excluded persons and, in some instances, by the perpetrator as well. By contrast, privacy protects behavior which is either morally neutral or valued by society as well as by the perpetrators. Secrecy is a more extreme form of denial of access to others than is privacy, because not only is access denied when secrecy is maintained, but the most successful secret occurs when knowledge of denial of access (the secret's very existence) is also withheld. However, the distinction between secrecy and privacy, since they may merge together, should not be overdrawn.

The commonsense meaning of privacy is indicated in the following phrases: "being alone"; "no one bothering me"; "controlling access to information"; and "controlling access to spaces."[10] *Webster's New Collegiate Dictionary* gives three meanings to *privacy*: "the quality or state of being apart from company or observation"; "seclusion"; and "secrecy." The meanings for the adjective *private* include "away from public view"; "secluded"; "not publicly or generally known"; "secret, confidential (a private matter)"; "not known or intended to be known publicly"; and "unsuitable for public use or display." All the above expressions or meanings make reference to separation from others through control over information, space or access, including simply being alone.

Westin defines privacy as "the claim of individuals, groups, or institutions to determine for themselves when, how and to what extent information about themselves is communicated to others."[11] Privacy, according to Irwin Altman, serves three functions: management of social interaction, establishment of plans and strategies for interacting with others, and development and maintenance of self-identity.[12] The privacy aspect of self-identity is related to personal autonomy and self-esteem:[13]

> Privacy mechanisms define the limits and boundaries of the self. When the permeability of these boundaries is under the control of a person, a sense of individuality develops. But it is not the inclusion or exclusion of others that is vital to self definition; it is the ability to regulate contact when desired. If I can control what is me and what is not me, if I can define what is me and not me, and if I can observe the limits and scope of my control, then I have taken major steps toward understanding and defining what I am. Thus, privacy mechanisms serve to help me define me.

The psychological aspect of privacy serves to maximize freedom of choice, to permit individuals to feel free to behave in a particular manner or to increase their range of options by removing certain classes of social constraints.[14] Westin has identified the following four psychological and physical dimensions of privacy:

(a) *Solitude*: This is the state where an individual is separated from the group and freed from the observations of others. It is the most complete state of privacy attainable although even here the subject's peace of mind may be intruded by physical stimuli, supernatural belief, or primordial psychological condition.

(b) *Intimacy*: This is the state where the individual is acting as part of a small group – the family, society, etc. Here corporate seclusion may be attained.

(c) *Anonymity*: This occurs where the individual, although doing public things in public places, finds freedom from identification and surveillance. Another form is the anonymous expression of views whereby the individual may publicly air his views but have his identity remain unknown.

(d) *Reserve*: This expresses the individual's need to withhold information, to create mental distance to protect his personality.[15]

These functional dimensions of privacy lead to the reinforcement of personal autonomy based on the belief in the uniqueness of the individual and his basic dignity and worth as a human being. Individuality and autonomy are concomitant factors. Privacy grants emotional release in various social roles. It also provides the opportunity for self-evaluation which is necessary to process daily experiences and to organize and cope with future experiences. This state of privacy is an insurance for limited and protected communication and for sharing confidences and intimacies on the basis of trust.[16] Neither affiliation nor disaffiliation is necessarily total. Accordingly, our identities are maintained by our ability to hold back as well as to affiliate.[17] This unity of opposites is the salt of a meaningful life; any total cohesion is the negation of life itself.

In a world of total social cohesion and mutual trust, privacy might be unnecessary. Where there is no antagonism, total happiness is achieved through a total harmony of parts. In *Brave New World*, Aldous Huxley portrayed one vision of that utopia. Harmony and happiness were abundant, but somehow an important dimension was missing. The individual had lost his chance to make a wrong decision in a programmed world, and with that loss went a loss of identity of individuality.[18]

Identity, individuality, and autonomy cumulatively involve questions of control or power related to privacy. Intrusion into privacy leads to erosion of one's power and to loss of control over self. The more the intruder knows about the intruded, the greater the balance of power shifts in favor of the former, the lesser the anonymity, autonomy, and control for the latter. So far as the intruded is concerned, Auguste Comte's dictum, "Know, in order to foretell; foretell, in order to control," begins to operate. The more the intruder knows about the intruded, the greater the control over the victim. From there on depersonalization and deindividuation sets in. The growth of human control techniques through surveillance or other forms offers the drawbacks of its success.[19] This is the reason why the word "privacy" is used in part as a synonym for political grievances about the use of information systems by institutions to enhance their power to the potential detriment of individuals, and in part for fears that information systems may be used to manipulate individuals or enforce conformity.[20]

Privacy as a constellation of values contains not only political significance but also ethical and aesthetic ingredients. The latter components assign value to privacy, not on the basis of utilitarian grounds, but on the basis of privacy *per se*; that is, privacy as an independent value by itself. This approach abhors the application of efficiency criteria to determine the value and dimension of privacy as a means to other ends.[21] This goes beyond the strategic consideration of privacy:

A fundamental distinction is that between what one might think of as "aesthetic" *versus* "strategic" considerations of privacy.

By *aesthetic* privacy, we mean the restriction of personal information as an end in itself. These are cases where disclosure is inherently embarrassing or distressing. The "instinctive" wish to shield from others actions of excretion, sexual intercourse, or profound emotion involves this form of privacy. If there is any parallel between information-restricting pursuits among humans and those of other animals, it undoubtedly has to do with what we term aesthetic privacy.

What we call *strategic* privacy is the restriction of personal information as a means to some other end. These are instances where privacy facilitates pursuit of other interests.[22]

On the basis of the preceding discussion, the concept and meaning of privacy is classified under the following categories:

I. Territorial Privacy.
II. Privacy of the Person/"Personal Space."
III. Aesthetic Privacy.
IV. Privacy in the Information Context.[23]

Territorial privacy claims in a spatial sense are related historically, legally, and conceptually to property. This involves a physical area within which a claim to be left in solitude and tranquility is advanced and is recognized. Since one's home is his castle, it is protected from trespass and nuisance.[24]

Privacy of one's physical person is deemed to be surrounded by a bubble or aura protecting one from intrusion or physical harassment. Intrusion implies physical as well as psychological components. This category is more abstract than the previous one because, unlike physical property, this "personal space" is not bounded by real walls or fences, but by legal norms and social values. What is protected is the integrity and dignity of the human person in a moral sense. Legal or constitutional guarantees of freedom of thought, expression, and movement, prohibition of physical assault, unreasonable search and seizure, and self-incrimination have a bearing on privacy.[25]

Aesthetic privacy, as a category, assigns an independent value to privacy *per se*. Here privacy is an end in itself, apart from strategic or utilitarian considerations.[26] This goes beyond the existing social and legal recognition of privacy.

Privacy in the information context is derived from the assumption that all information about a person is in a fundamental way his own, for him to communicate or retain for himself as he sees fit.[27] Information about a person may be collected by another person or an organization through voluntary self-disclosure or through other means. In addition to these methods, this category deals with data management involving the following three aspects:[28]

(a) *Privacy*: the question of what personal information it is relevant and proper for an organization to collect or store at all.
(b) *Confidentiality*: the question of how information about individuals should be distributed within the organization and when it should be released to outsiders.
(c) *Individual access*: the question of whether individuals should be able to learn what information about them is being maintained in the data system, and have the opportunity to inspect, correct, or contest such data.

These questions have gained social urgency and significance in the recent past because of the superimposition of an accelerating information technology over the already existing network of huge and complex formal organizations in public and private sectors. Hence privacy should be examined in its technological context.

PRIVACY AND THE TECHNOLOGICAL CONTEXT

Technology consists of two components: one is social technology, such as a modern corporation or bureaucracy, and the other is physical technology, such as computers or microelectronics technology. As for the privacy questions raised previously, the information technology – given the rate of innovations, the scale, the pervasiveness and the very magnitude of change wrought by it – in its interaction with complex organizations has a triggering role. The physical and social technologies, depending upon the circumstances, may act as concomitant variables or in a cause-effect manner in introducing changes in the system. In either way, our concern here is the cumulative impact of the evolving quantitative changes leading to qualitative changes in privacy issues:

Concrete experiences with privacy are transmitted through the institutional fabric of a society. Privacy options are a function of the ecological and physical properties of the environmental settings that circumscribe human behavior. The available technology (doors, indoor plumbing, personality tests, computers), the types of tasks required, the numbers of people and their relation to one another, and specific ritualistic activities reflect as well as influence the kinds of activities and environments that are considered private.[29]

Bureaucratization and informatization of society are two interrelated processes in the sense that the triumph of information technology finds a striking parallel in the growing role of formal organizations, of bureaucracies, in society.[30] The bureaucratization of society is the point from which a macrosociological analysis of the societal implications of automated data processing (ADP) could take its start. Both are rooted in the apparent superiority of formalized structures for achieving particular ends. The unorganized, unregulated, unstructured, and personal segments of society could reach a vanishing point. It is in this sense that information technology, in conjunction with public and private bureaucracies, makes individual freedom and privacy more vulnerable. A major source of technological vulnerability lies in the sheer size and complexity of many automated information systems which aim at "perfect administration":

ADP is important to bureaucracies because it is the embodiment of rational administrative behaviour directed at the efficient attainment of organisational goals. It is a powerful aid at the service of the bureaucratic mode of coping with social problems of all kinds. Computers tend to reinforce the secular process of bureaucratization of the society as they enable bureaucratic organisations to respond to rising demands and increased societal complexity. In doing so, they contribute to making individuals, as well as society at large, still more dependent on the bureaucratic organisations that structure social intercourse and cater to many human needs.[31]

According to James B. Rule,[32] the organizations concerned have hardly developed their present appetite for personal data as an end in itself. Rather this appetite has developed in order to satisfy demands for authoritative actions concerning people depicted in the records. People expect certain organizations to deal intelligently with a heterogeneous array of individuals. In every case the organization is expected to render to each person his or her "due," that is, the "correct" form of bureaucratic action, in the light of all relevant information on that person's past history and current status, whether it is income taxation, consumer credit, law-enforcement, or personnel administration. In this sense, there is nothing new about the appetite of organizations for personal information. But what is new is the synchronization of bureaucratization and informatization through the new technology. "Organizations have invaded people's privacy with steel file cabinets and manila folders for years. But computer systems with remote access have intensified both the problem and public concern."[33]

LAW AND PRIVACY IN CANADA

Historically, the concept of a constitutional democracy has presupposed individual privacy in its widest sense.[34] With the advent of datamation as an industry, it had become clear even by the 1960s that the political system that could not have been built without privacy cannot now endure without its protection. Protection of privacy through common law principles had been found to be inadequate given the lag between and among economics, law, and technology:

The legal system has been a passive bystander to structural and economic changes which the rise in technology has brought to our society. The legal system has been a co-conspirator in conserving the existing system. Law has allowed a legal-economic framework to be created by which the fruits of technology are monopolized and the invisible "costs" are disseminated. The legal system, by holding corporate rights to be co-equal with individual rights, have influenced the evolution of technology. The legal system has placed no restraints on the rate of economic surplus developed by technology at great cost to the community and no constraints on the distribution or utilization of this abundance. Technology has unfettered legal and economic freedom without manifesting social utility.[35]

The inadequacy of the protection provided under the established heads of tort, liability, judicial conservatism, and a dearth of cases, combined with the increasing threats to privacy by modern technology, led the provincial legislatures of British Columbia, Manitoba, and Saskatchewan to enact a tort of "invading privacy."[36] British Columbia was the first jurisdiction in Canada and the Commonwealth to inquire into the matter through a Commission and to enact the first Privacy Act in 1968.[37] Since that time a few other jurisdictions have enacted similar privacy legislation.[38]

The effectiveness of these Privacy Acts has been assessed somewhat negatively by Burns – labelling this "significant development" a "non-development"[39] – and by Ryan as a "well intentioned dead letter."[40] P. H. Osborne has identified the following four categories of reasons for this development of non-development: the uncertainty and unpredictability in the application of the acts, the heavy burden on the plaintiff to prove actionable invasion of privacy, the broad and nebulous scope of the defenses, and the denial of access to the lower courts to litigate claims.[41] In his assessment, to draft privacy legislation in the static common-law manner would be to make the legislation an anachronism the moment it is passed.[42]

Apart from these developments in privacy legislation, a series of regulations regarding the personal reporting industry has emerged during the 1970s.

REGULATION REGARDING THE PERSONAL REPORTING INDUSTRY

The personal investigation and reporting industry consists of "credit bureaus," "credit reporters," "commercial reporters," "reporting agencies," and "security and investigation agencies."[43] Their function is personal information collection, storage, and dissemination.[44] Potential employers and insurers lean heavily on this industry. The reporting agencies in the U.S. maintain and issue more than a hundred million personal files every year, and well over ten million personal reports are issued annually in Canada.[45]

The magnitude of the information transacted and the real and potential errors in data collection, transmission, and use prompted the U.S. Congress to pass the first major legislation in 1970 to deal with the problem.[46] Since then, the following Canadian jurisdictions have enacted measures regulating the reporting industry: Manitoba, Saskatchewan, Ontario, Nova Scotia, Newfoundland, British Columbia, and Prince Edward Island.[47]

Though these legislative acts are a step forward, they are found to suffer from the following problems: narrow scope, wide exceptions; lack of ample powers of inspection, investigation, conciliation, correction of files and reports, and effective enforcement; ineffective and inadequate notice to subjects upon whom information is collected; inadequate emphasis on information storage practices; and the lack of uniformity between jurisdictions.[48] The model Information Reporting Act released in 1977 by the Uniform Law Conference has suggested a set of valuable guidelines for further legislative changes.[49]

PRIVACY AND PERSONNEL ADMINISTRATION

All these pieces of legislation regulate matters of privacy from different angles and for different purposes. But privacy of persons as prospective or current employees is covered only in a marginal fashion. The reason for this marginality is the special nature of the relationship between employers and employees. The area of employee privacy in personnel administration as a field is still underdeveloped, in spite of the fact that more than ninety percent of the labor force consists of "employees." A recent Canadian text on personnel management disposes of the privacy issue of employees in the following two sentences, under the heading 'Selected Challenges Facing Personnel Management in the 1980s and 1990': "With the growth of computerized personnel information systems, companies maintain ever-growing data banks on employees. The need to ensure privacy of employees against abuses of this information may cause greater attention to be paid to the privacy of employee records."[50] An American text on the same subject states that all the specific details on how exactly the privacy laws will affect personnel records and personnel information systems have not yet been determined.[51]

A recent American study on privacy in personnel administration[52] has rectified the hitherto existing paucity of attention to this area. The study report opens its policy analysis with the judgment that measures do need to be taken to assure that citizens' rights are effectively provided for in the use of personal data in the employment process. While this is true for both manual and automated record systems, the study emphasizes that it is especially important where automated data systems contain substantial amounts of sensitive information that is capable of rapid access and extensive dissemination.[53] The report covers employers in the governmental, business, and non-profit settings. Based on public discussion, organizational policies, and legal enactments during the past decade, the report identified the following six principles of policy:[54]

(1) Decisions about an individual's rights, benefits, and opportunities in society should not be made by organizations on the basis of secret files, or of record-based procedures about which individuals are not informed.

(2) Only information relevant to the organization's legitimate purposes should be collected and stored, and the definition of relevance must respect both guarantees of privacy and legislative prohibitions against making improper racial, sexual, cultural, and similar discriminatory decisions.

(3) Managers of a data system should take reasonable steps to insure that the records they keep are accurate, timely, and complete, as measured by the kinds of uses made of the data and the social impact of their use.

(4) Detailed rules of confidentiality should govern who within the organization maintaining the data system has access to a record, and this should be based on a need-to-know principle.

(5) Disclosure of personal data outside the organization that collected the data should be made only with the informed and voluntary consent of the individual, obtained at the time of collection or by subsequent query, or under a constitutionally valid legal order.

(6) An individual should have a right to see his or her record, and have an effective procedure for contesting the accuracy, timeliness, and pertinency of the information in it. There may be some exceptions to this right of inspection, as in the interests of protecting confidential law enforcement sources, but these should be rare.

These six principles of policy have the following three broader goals: *to minimize intrusiveness*: to create a proper balance between what an individual is expected to divulge to a record-keeping organization and what he seeks in return; *to maximize fairness*: to open up record-keeping operations in ways that will minimize the extent to which recorded information about an individual is itself a source of unfairness in any decision about him made on the basis of it; and *to create legitimate, enforceable expectations of confidentiality*: to create and define obligations with respect to the uses and disclosures that will be made of recorded information about an individual.[55]

To achieve these broader goals in personnel data systems, three dimensions of the employment situation – namely, pre-employment, employment, and post-employment – have to be examined in terms of personnel data gathering, storage, handling, use, access, and dissemination.

At the pre-employment stage of data gathering, the issues involved are the method of gathering the data, the relevancy and accuracy of the data, and information *wanted* as opposed to what information is *really needed* by the various governmental, business, and non-profit organizations. The primary source is the collection of data, through the application form and interview with the applicant; the applicant's self-disclosure is the primary source. Various human rights codes prohibit collection of certain information through the application form as well as through questions at personal employment interviews.[56]

"File" agencies, such as a credit reporting agency, and "investigative" agencies are the two main outside sources which are used by law enforcement agencies, insurance companies, social welfare agencies, regulatory agencies, merchandising houses and employers. Most information systems are local in a formal sense, but through information interchange agreements they are able to conduct operations on a national and international scale.[57] Data collected through "filing" and investigative agencies are used by employers in hiring as well as for monitoring, promoting, and firing employees.[58] Investigators who collect information are found to be among the lowest-paid, least-trained members of their particular organizations.[59] When they make inquiries among the friends and neighbors of a subject, the final dossier is vulnerable to hearsay and gossip based on prejudice and malice. Thus the manner and methods adopted by employers in collecting data may not only contribute to the inaccuracy and irrelevancy but also to the intrusion into personal privacy. If the job is so sensitive as to require the services of an investigative agency, the subject should be informed of such investigation as well as the nature and purpose of investigation.

A case in point is *Gillett v. Nissen Volkswagen Ltd. et al.*[60] Mr. Gillett, with wide experience in the automobile business, applied for the position of regional sales manager with a Calgary automobile dealer. He was a leading candidate for the position. The dealer engaged the Retail Credit Company of Canada Ltd. to investigate the various applicants. Among other things, this agency had an interview with a former employer of Gillett and made the following statement about this applicant:

He was let go from the firm as his honesty was questioned. The company was satisfied that there was [sic] some irregularities in his handling of company money and for this reason he would not be eligible for rehire. No charges were laid and they did not ask for restitution. They declined to go into any further details and refused to make any further comment beyond this point.[61]

It was established in the proceedings that the above statements were completely false. Gillett was not let go. He resigned. He had never handled any company money. His honesty had never been questioned. The statements had been made maliciously by the former employer because Gillett had left his employ to work for a competitor. The reporting agency did not have any corroborating evidence to support these damaging allegations. Gillett won a defamation action against both the former employer and the reporting agency but did not get the job.

Given the fact that the reporting agencies in the U.S. maintain and use a hundred million files, and in Canada over ten million files,[62] Dale Gibson has assessed the seriousness of the problem in these words:

The reporting industry asserts that it exercises great care in amassing, storing and transmitting personal information. Generally speaking, this is probably true. But given the number of information transactions that take place, the fact that each one involves the participation of several persons, and the human propensity for error, exaggeration, misunderstanding and spite, it is inevitable that substantial numbers of errors occur. An official of an organization to which most North American agencies belong estimated, a few years ago, that "complaints of mistaken identity" are received "in less than 1 percent of all the reports." Bearing in mind that errors which do not come to light cannot be complained about, and that "mistaken identity" is far from the only type of error that can be made, it would appear that the overall error rate must be much higher than that. However, even if we were to accept "less than 1 percent" (say 0.5 percent) as the total rate, more than 50 000 errors annually would be indicated for Canada alone. If only 10 percent – 5000 – of those errors produced major adverse consequences for Canadians, a serious problem would exist. In all probability the problem is many times greater than that.[63]

In selection and placement, a variety of instruments, such as measures of aptitude, abilities, personality, and interests, are in use. Since individuals transmit a great deal of information about themselves to the employer through these measures, issues of privacy should be a concern not only to the subject but also to personnel psychologists and managers.[64] It has been asserted that despite the legislative and public concern with invasion of privacy, personnel psychologists have not responded to these concerns in any major way.[65] It is the employers' right to know which is emphasized in the application of these instruments and individual privacy from the employee's perspective is generally ignored. Personal history data should not be collected through hidden exploration of the psyche; such psychological espionage could be more intrusive than the ordinary type of surveillance. The subject should be informed of the nature and purpose of tests and also of storage and use of such data. Data collected for one purpose should not be used for an entirely different purpose.

Individuals should be made aware of the content of their files at the employment stage. They should have access to their files periodically so that outmoded information may be eliminated and inaccuracies corrected. When the computerized system goes beyond the payroll type, clear cut policy must be formulated regarding data storage, handling, use, dissemination, and third-party access as well as disposal of personnel data after termination.

Given international unions, governmental co-operation between Canada and the U.S.A., and the dominant role played by multinational corporations, there should be a clear cut policy regarding transnational transfer of personal data. The O.E.C.D., which has conducted several studies into the area, concluded that it is governments that can do the most to help since governments are the most assiduous collectors, evaluators, and transmitters of information and are thereby in the best position to enforce high standards and set good examples.[66] There are contradictory claims and rights regarding information; for instance, the public's right to know, as embodied in freedom of information enactments, is partly in conflict with privacy enactments. In this conflict individual privacy should take precedence over the freedom of information. This task demands a sensitive balancing in the context of particular situations.[67] The following observation on personnel security is worth noting.

Anyone entrusted with access rights to data is a potential security leak for that data. The most direct method of access for an intruder is often through a person in a position of trust. This problem is well known, and several methods are used to decrease the probability of a betrayal of trust. Amongst these are personnel investigation to determine trustworthiness, and the imposition of penalties for breach of security. Above all, the opportunity for accidental disclosure should be minimized by having precise security procedures with regard to labelling of sensitive data, locking of file cabinets, etc. The problems here are identical with those of security of data in a manual system.[68]

PRIVACY: VALUES *VIS-À-VIS* EFFICIENCY

The growth of the datamation industry has brought with it a new value. This value equates information with wisdom: collecting information for its own sake. The quantitative form lends an aura of rigor and infallibility. "Inefficiency" in the past has helped to shelter privacy. The enthronement of efficiency criteria supported by datamation accelerates the invasion of privacy. Privacy is not considered as a social and individual good; it is not cherished as an independent value *per se.* When more and more efficiency criteria are applied, the value of privacy is determined on utilitarian grounds. Under these criteria, there is no *prima facie* right to privacy.

By fostering an interpretation of privacy protection in terms of utilitarianism, efficiency criteria, and procedural safeguards, both the critics and the defenders of surveillance practices could avoid the really difficult and painful questions: How much surveillance is a desirable thing? How far should the development or bureaucratic monitoring of otherwise private affairs be extended?[69] As it is, the privacy planners do not engage themselves in resolving these difficult but fundamental questions; instead, they look at privacy issues in terms of the efficiency criterion.

By this criterion, privacy is deemed protected if three conditions are met in managing personal data: (1) that the data be kept accurate, complete, up-to-date, and subject to review and correction by the persons concerned; (2) that the uses of filed data proceed according to rules of due process that data subjects can know and, if necessary, invoke; and (3) that the organizations collecting and using personal data do so only insofar as necessary to attain their appropriate organizational goals. Under these principles, organizations can claim to protect the privacy of the persons with whom they deal, even as they accumulate more and more data on those persons and greater and greater power over their lives. It would be difficult to imagine a more advantageous interpretation of privacy protection, from the standpoint of surveillance organizations. . . .

So long as the efficiency criterion continues to guide the development of these systems, their attentions will continue to spread over larger and larger areas of what had been private experience. At some point, nearly everyone would acknowledge, such extension passes the point of moral or aesthetic acceptability.[70]

We have not yet reached that point of unacceptability but are tending toward it. We have two notions of paradoxical conventional wisdom; one is that "people must reap as they sow," and the other one asserts that "people deserve a fresh start or a second chance." We seem to be developing a personal and personnel data system which is not only "unforgetting" and "unforgiving" but also has all the deceptive appearances of accuracy, authenticity, and infallibility. The two components of the conventional wisdom are not necessarily mutually exclusive. There is a synthesis between them. That should be the foundation for a data system policy to enhance personal dignity, autonomy, and privacy.

POLICY CONSIDERATIONS

Hitherto we have examined privacy under six headings: privacy: concept and meaning; privacy and the technological context; law and privacy in Canada; regulations regarding the personal reporting industry; privacy and personnel administration; and privacy: values *vis-à-vis* efficiency. The question now is what is to be done about the various issues raised.

One easy alternative is to treat privacy as a non-issue, so that no action is required. Partial justification for this stand is that there are enough regulations already in existence, and the various governmental, business, and non-profit organizations are in a position to regulate themselves accordingly. But the evidence examined in this exploratory study indicates otherwise. The need for creating social awareness of the problem in general and the need for action particularly in personnel data system development and management is pressing.

The most preferred choice is to encourage voluntary action on the basis of needs, goals, experiences, and scale of different organizations. The emphasis in this option is on self-analysis, self-initiative, self-regulation, and self-enforcement of privacy policies. One-third of the labor force is organized. The collective bargaining system to a limited extent is amenable to such action in the organized sector.[71]

But collective bargaining is not an all-purpose institution to find an effective remedy for the privacy issue. It has its priorities and privacy may not be one of them, given the economic and political climate. A recent Health and Welfare Department questionnaire, known as Evalu Life, inquiring about employees' smoking and drinking habits, suicidal tendencies, sex and rectal disorders, is an example.[72] According to the Canadian Federation of Communication Workers, the questionnaire represents a flagrant intrusion into the personal lives of thousands of workers.

The department's defense is that it is to be used by the consent of the employees. Is it reasonable and proper to regard consent as a defense to an invasion of privacy irrespective of the circumstances and purposes of the invasion? In many areas it is becoming increasingly necessary to "volunteer" information about oneself.[73] In some cases there may be a subtle, but unprovable, element of duress or undue influence to vitiate the consent. Other than this, there are problems related to expressed consent and implied consent. What is the scope of these consents? What is the duration of the consent? Is it revocable? If so, when and how? When does a consent become contractual consent? These questions are not amenable to resolution through collective bargaining. Even if they were, more than two-thirds of the labor force are not unionized.

There is a general institutional inertia on the part of various organizations to do something good for its own sake. When an action involves cost, it is postponed as long as possible. To what extent enlightened self-interest in conjunction with public interest would induce self-regulation and enforcement of privacy matters is debatable, but such action is not improbable. Bank of America[74] and IBM[75] are examples of such self-regulation.

If the institutional inertia, as indicated above, is prolonged, then some kind of legislative action on employee privacy, such as the *Michigan Employee Right To Know*, has to be contemplated.[76]

This law covers all enterprises with four or more employees and grants employees the right to see any information in their personnel files used in determining employment, promotion, transfer, additional compensation, or disciplinary action.[77] Employees have the right to put in their files a written rebuttal of any information felt to be incorrect or unfair. The law also prohibits employers from including in personnel files any information relating to an individual's associations, political activities, publications, or communications on non-employment activities unless the employee gives written permission. The employer is also required to give his employees written notification whenever disciplinary reports or letters of reprimand are disclosed to a third party, unless the employee waives this requirement. Any derogatory information more than four years old may not be sent to third parties.

No state agency is charged with administering this law, but an employee can sue to recover actual damages, a $200 penalty, court costs, and reasonable attorney's fees.

Policy developments in Europe should also be noted here. In Europe, North America, and Japan, home information services, such as videotex services, are developing rapidly.[78] These services consist of linking a suitably modified television set equipped with a touch-tone pad to remote data bases using normal telephone lines. In assessing this development, H. P. Gassman has raised the following questions regarding the first generation laws on privacy and also indicated the desirable direction of policy change:

The central question concerning privacy protection is how much personal data will be handled, stored, and transmitted by these systems? Will the existing first-generation laws on privacy protection curb the rapid growth and dissemination of personal information in electronic form, or will the market forces be so strong that collection and dissemination of such data will continue to increase at the same pace as they have done so far, but adhering to the new provisions and safeguards as enunciated in privacy protection legislation? . . .

It is probable that we might see a generalization of second-generation privacy laws – those covering not only physical persons (private lives), but also extending to legal persons (business lives). At present, at least in Europe, there exists a clear trend toward such; legislation passed in 1978 (Norway, Denmark, Austria) and 1979 (Luxembourg) contains provisions on legal persons.[79]

CONCLUSION

This study has explored the privacy policy developments primarily in Canada and the U.S. and secondarily in some of the European countries in the light of the developments in information technology. The existing first generation laws on privacy protection have been found to be lacking in substance and scope as well as weak in enforcement. These developments in the late sixties and early seventies have become a "non-development" and "a well intentioned dead letter." The gap between and among economics, law, and the new technology is widening. The pure and simple utilitarian approach to privacy of persons and personnel is found to be outmoded. There is a pressing need for creating a *prima facie* right to privacy. As it is, the burden of proof on the victims is too heavy, unfair, and unwarranted. A *prima facie* right to privacy will not only make privacy a social and individual good but also shift part of the burden of proof to the violator to justify the invasion of privacy.

The central policy question is: What is to be done? Will voluntarism work? There is a reasonable doubt on the efficacy of voluntarism to resolve issues related to privacy. The author's bias is in favor of legislative action along the line suggested by the model Information Reporting Act released in 1977 by the Canadian Uniform Law Conference.[80]

Privacy is a constellation of values, claims, and interests in a universe of concurring and competing values, of allied and adverse interests. Institutional wisdom lies in balancing the conflicting interests and rights in applying information technology. Technology is neutral in its abstraction but not necessarily so in its application. A *laissez-faire* approach to it may amount to technological determinism. Information technology is vital to our economy. It can facilitate the coming of a new society with a qualitative difference, but it cannot construct such a society on its own initiative. That is the function of human choice made in a humane way. Now is the time to make such a conscious choice.

University of Regina

NOTES

[1] Department of Communication/Department of Justice, *Privacy and Computers* (Task Force Report; Ottawa: Information Canada, 1971), p. 11. Hereafter, this source is cited as *Privacy and Computers*.

[2] R. S. Laufer and M. Wolfe, 'Privacy as a Concept and a Social Issue: A Multidimensional Development Theory', *Journal of Social Issues*, **33**, no. 3 (1977): 25.

[3] *Ibid.*

[4] See in general A. Westin, *Privacy and Freedom* (New York: Atheneum, 1967). This is one of the earliest and most influential works on this subject. Also, see Irwin Altman, 'Privacy Regulation: Culturally Universal or Culturally Specific?', *Journal of Social Issues*, **33**, no. 3 (1977): 66–84.

[5] S. T. Margulis, 'Conceptions of Privacy: Current Status and Next Steps', *Journal of Social Issues*, **33**, no. 3 (1977): 12.

[6] Altman, *op. cit.* (note 4, above), p. 66.

[7] Westin, *op. cit.* (note 4, above), pp. 8–9. Privacy as a social reality has been traced back to the Bible, Socrates, Plato, Thomas More, and Locke, by Konvitz, 'Privacy and the Law: A Philosophical Prelude', *Law and Contemporary Problems*, **31** (1966): 273–275.

[8] C. Warren and B. Laslett, 'Privacy and Secrecy: A Conceptual Comparison', *Journal of Social Issues*, **33**, no. 3 (1977): 43.

[9] *Ibid.*, p. 44. The state and certain components of it have limited the legal right to secrecy. Limits of this right are based on the constitution and other pertinent legislation in each democratic state. Further, "trade secret" is a class by itself regulated through a different set of principles. For more details on this subject, see H. J. Glasbeck, 'Limitations on the Action of Breach of Confidence', in Dale Gibson ed., *Aspects of Privacy Law* (London: Butterworths, 1980), pp. 228–244.

[10] Wolf and Laufer, cited in Margulis, *op. cit.* (note 5, above), p. 7.

[11] Westin, *op. cit.* (note 4, above), p. 7.

[12] Altman, *op. cit.* (note 4, above), p. 68.

[13] Altman, *The Environment and Social Behavior: Privacy, Personal Space, Territory, and Crowding*, (Monterey, CA.: Brooks/Cole, 1975), p. 50.

[14] Proshansky, Ittelson, and Rivlin, 'Freedom of Choice and Behavior in a Physical Setting', in *Environmental Psychology: Man and His Physical Setting* (New York: Holt, Rinehart & Winston, 1970), p. 173.

[15] 'Science, Privacy and Freedom: Issues and Proposals for the 1970's', *Columbia Law Review*, **66** (1966): 1020–1021.

[16] *Ibid.*, pp. 1022–1028.

[17] B. Schwartz, 'The Social Psychology of Privacy', *American Journal of Sociology* (May 1968): 752.

[18] *Privacy and Computers* (note 1, above), p. 18.

[19] J. B. Rule *et al.*, 'Preserving Individual Autonomy in an Information Oriented Society', in L. J. Hoffman, ed., *Computers and Privacy in the Next Decade* (New York: Academic Press, 1980), p. 84.

[20] *Privacy and Computers*, p. 183.

[21] J. B. Rule *et al.*, *op. cit.* (note 19, above), p. 74. Also see J. B. Rule, *et al.*, *The Politics of Privacy*, (New York: New American Library, 1980), pp. 22–23.

[22] *The Politics of Privacy*, *ibid.*

[23] These categories are not discrete but are based on conceptual convenience. See *Privacy and Computers* (note 1, above), pp. 12–14. This classification is generally in tune with the following legal aspect of privacy:

Tort: Person	Tort: Property	Other Legal Aspects
Safety	Nuisance	Criminal Code: Watching-Besetting
Repute	Trespass	Quebec Civil Code, 1968
		Privacy Act
		Human Rights Code
		Charter of Rights
		Contract Law
		Breach of Confidentiality

[24] *Privacy and Computers* (note 1, above), p. 13.

[25] *Ibid.* The new Charter of Rights in the Canadian Constitution, though untested yet, has a bearing on the concept of privacy; see sections 1, 6, 7, 8, 9, 10, and 11.

[26] Rule, *Politics of Privacy* (note 21, above), p. 22.

[27] *Privacy and Computers*, (note 1, above), p. 13.

[28] Alan F. Westin, *Computers, Personnel Administration, and Citizen Rights* (NBS Special Publication, 500–50; Washington, D.C.: U.S. Department of Commerce, National Bureau of Standards, 1979), p. 10.

[29] Laufer and Wolfe, *op. cit.* (note 2, above), pp. 28–29. For the interaction between physical and social technologies, see S. Coates, *The Office of the Future* (Ottawa: Department of Communications, 1982).

[30] K. Lenk, 'Information Technology and Society', in G. Friedricks and A. Schaff, eds., *Microelectronics and Society: For Better or for Worse* (A Report to the Club of Rome; Oxford: Pergamon Press, 1982), p. 304.

[31] *Ibid.*, p. 305.

[32] Rule *et al.*, *op. cit.* (note 19, above), p. 66.

[33] Statement by Frank T. Cary, Chairman of the Board and Chief Executive Officer of IBM, 'IBM's Guidelines for Employee Privacy', *Harvard Business Review* (September–October, 1976): 82.

[34] E. F. Ryan, 'Privacy, Orthodoxy and Democracy', *The Canadian Bar Review*, **51** (1973): 85.

[35] J. S. Grafstein, 'Law and Technology: A Technological Bill of Rights', *The Canadian Bar Review*, **51** (1973): 230. Also see A. E. Gotlieb (Deputy Minister, Communications, Ottawa), 'Some Social and Legal Implications of New Technology: The Impact of Communications and Computers', *The Canadian Bar Review*, **51** (1973): 247–255. For an excellent analysis of common law principles as applied to privacy, see P. Burns, 'The Law

and Privacy: The Canadian Experience', *The Canadian Bar Review*, **54** (1976): 12–24. For common law and other legal principles related to privacy, see note 23, above.

[36] P. H. Osborne, 'The Privacy Acts of British Columbia, Manitoba and Saskatchewan', in D. Gibson, ed., *Aspects of Privacy Laws*, *op. cit.* (note 9, above), p. 81.

[37] Sargent, Report of the Commission of Inquiry into Invasion of Privacy, August 9, 1967 (B.C.). The British Columbia Privacy Act, S.B.C., 1968, c.39.

[38] The Manitoba Privacy Act, S.M., 1970, c.74; The Saskatchewan Privacy Act, S.S., 1974, c.80; The Federal Statute, the Protection of Privacy Act, S.C., 1973–74, c.50. Recently the Federal Bill C-43 has combined the Access to Information Act and the Privacy Act in one piece of legislation. For a discussion of this combined act, see B. Land, 'Understanding the New Information Act', *CAUT Bulletin*, May 1983. Ontario Consumer Reporting Act, S.O., 1973, c.97; Nova Scotia Consumer Reporting Act, S.N.S., 1973, c.4; for developments in Quebec, see Glenn, 'Civil Responsibility – Right to Privacy in Quebec', *Canadian Bar Review*, **52** (1974)): 297–305. There are some variations between and among the various privacy acts in Canadian jurisdictions. Due to limitation of space, these variations are not discussed. For a comparative analysis of these acts, see P. Burns, *op. cit.*, note 35, above. B.C., Ontario, Nova Scotia, and Saskatchewan privacy acts are very similar.

[39] 'The Law and Privacy: The Canadian Experience' (note 35, above), p. 33.

[40] Ontario Law Reform Commission, 'Report on Protection of Privacy in Ontario' (1968), p. 68.

[41] *Op. cit.* (note 36, above), p. 108.

[42] *Ibid.*, p. 110.

[43] Regarding the last category, see C. D. Shearing and P. C. Stenning, *Private Security and Private Justice: The Challenge of the 80's* (The Institute for Research on Public Policy, 1983). The authors suggest that the policy makers must not only be concerned with the manner in which information is collected, but also the use to which such information is put once it has been collected.

[44] Dale Gibson, 'Regulating the Personal Reporting Industry', in Dale Gibson, ed., *Aspects of Privacy Law*, *op. cit.* (note 9, above). Also see Gibson and Sharp, *Privacy and Commercial Reporting Agencies* (Winnipeg: Legal Research Institute, University of Manitoba, 1968). Other studies similar to the Manitoba report: Department of Communications, *Conference on Computers: Privacy and Freedom of Information* (Queen's University, 1970); Sharp, *Credit Reporting and Privacy: The Law in Canada and the U.S.A.* (1970); and the Federal Report, *Privacy and Computers*, *op. cit.* (note 1, above). For more details, see P. Burns, *op. cit.* (note 35, above), pp. 39–41.

[45] Dale Gibson, 'Regulating the Personal Reporting Industry', *op. cit.* (note 44), p. 113.

[46] *Fair Credit Reporting Act*, Public Law 90–321, 1970; 15 U.S.C.A. 1681. Amended by Public Law 95–598, 1978.

[47] These enactments respectively are: C.C.S.M., c. P33; S.S. 1972, c. 23; S.O. 1973, c. 97; S.N.S. 1973, c. 4; S.N. 1973, c. 76; S.B.C. 1973 (2d), c. 139; and S.P.E.I. 1974, c. 67. On Quebec, see H. P. Glenn, 'The Right to Privacy in Quebec Law', in D. Gibson, *Aspects of Privacy Law*, *op. cit.* (note 9, above).

[48] See L. D. Frenzel, 'Fair Credit Reporting Act: The Case for Revision', *Loyola (L.A.) Law Review*, **10** (1977): 409; S. P. Whelan, 'A Comparison and Criticism of the Saskatchewan Credit Reporting Agencies Act', *Saskatchewan Law Review*, **41** (1976–1977): 215.

[49] For more details, see note 45, above.

[50] W. B. Werther, Jr., *et al.*, *Canadian Personnel Management and Human Resources* (Toronto: McGraw-Hill Ryerson, 1982), p. 467.

[51] R. L. Mathis and J. H. Jackson, *Personnel: Contemporary Perspectives and Applications* (2d ed.; St. Paul: West, 1979), p. 443.

[52] A. F. Westin, *op. cit.* (note 28, above).

[53] *Ibid.*, p. xvii.

[54] *Ibid.*, p. xviii.

[55] Privacy Protection Study Commission, *Personal Privacy in an Information Society* (Washington, D.C.: U.S. Government Printing Office, 1977), pp. 14–15.

[56] For more details, see the guidelines issued by human rights commissions in the appropriate jurisdictions. For example: Ontario Human Rights Commission, *Employment Applications Forms and Interviews* (1982); Saskatchewan Human Rights Commission, *A Guide for Employers Regarding Employment Applications Forms and Interviews under the Saskatchewan Human Rights Legislation* (no date).

[57] For more details, see *Privacy and Computers, op. cit.* (note 1, above), pp. 34–35 and 54–67.

[58] *Ibid.*, p. 54.

[59] *Ibid.*, p. 113.

[60] (1976), 58 D:L.R. (3d), 104 (Alta.). For a recent incident, see 'Detective Agency Is Banned by Credit Bureau over Search', *Globe and Mail*, January 2, 1984.

[61] P. 107.

[62] See note 44, above.

[63] *Ibid*, p. 113. Also see J. K. Fedorowicz and G. Lucas, 'Groping for a Door Marked Data', *Globe and Mail*, December 12, 1983, p. 7; and Andrew Pollack, 'The Right to Privacy in Computerized World', *Globe and Mail*, January 2, 1984, p. 7.

[64] V. E. Schein, 'Individual Privacy and Personnel Psychology: The Need for a Broader Perspective', *Journal of Social Issues*, **33**, no. 3 (1977): 155.

[65] *Ibid.*, p. 157. For a legislative model, see *State of New York Act to Amend The Labor Law*, 6140, 1977–1978, March 1, 1977; the act amends the labor law in relation to the prohibition of psychological stress evaluator examinations.

[66] *Privacy and Computers, op. cit.* (note 1, above), p. 166.

[67] *Ibid.*, p. 121.

[68] *Ibid.*, p. 107.

[69] J. B. Rule *et al.*, *op. cit.* (note 19, above), p. 74.

[70] *Ibid.*

[71] For a model collective agreement, see 'Data Protection Agreement between Lucas Industries (UK) and Unions APEX and ACTSS', *European Industrial Relations Review*, **110** (March 1983): 20.

[72] See 'Union Hits Roof over Health Questionnaire', *Globe and Mail*, January 26, 1984, p. 12. The Department distributes the form, containing 25 questions, as a public service to about 300 corporate and professional organizations across the country.

[73] See P. H. Osborne, 'The Privacy Acts', in Gibson, ed., *Aspects of Privacy Law, op. cit.* (note 9, above), pp. 90–94.

[74] *Computers, Personnel Administration, and Citizen Rights, op. cit.* (note 28, above), pp. 203–225.

[75] For I.B.M.'s guidelines to employee privacy, see *Harvard Business Review, op. cit.* (note 33, above).

[76] *Michigan Employee Right To Know Act*, Act No. 397. Public Acts of 1978. In effect January 1, 1979.

[77] See A. F. Westin, 'Problems of Employee Privacy Still Trouble Management', in A. F. Westin and S. Salisbury, eds., *Individual Rights in the Corporations: A Reader on Employee Rights* (New York: Pantheon Books, 1980), pp. 245–256.

[78] For an excellent analysis and overview of the causes and consequences of the renaissance of homeworking in many industries in Western Europe, North America, and Australia, see Carla Lipsig-Mumme, 'The Renaissance of Homeworking in Developed Economies', *Relations Industrielles*, **38**, no. 3 (1983): 545–566.

[79] 'Privacy Implications of Transborder Data Flows: Outlook for the 1980s', in L. J. Hoffman, ed., *Computers and Privacy in the Next Decade* (New York: Academic Press, 1980), p. 111. For more details regarding the interaction between law and computer technology in specific countries, see the various issues of *Law and Computer Technology* since 1979.

[80] Justice Minister, Province of Quebec, intends to introduce legislation to control the use of personal information in data banks for the following reasons: "Because of accelerated computerization in all sectors of society, it has become an urgent necessity [for the Government] to intervene in this field to avoid inevitable abuses." Also, "Computerization of personal information leads to its concentration and considerably increases [the ease of] distribution." And finally, "It is therefore time to question certain methods of gathering information, storing it and managing it, as well as to recognize certain abusive uses and especially the importance of existing law in this area." *Globe and Mail*, January 17, 1984, p. 5.

JOSEPH MARGOLIS

HISTORY, NATURE, AND TECHNOLOGY

I

In the middle of an extremely clever but temperate attempt to distinguish a truly modern conception of history – *not* in the thrall of ancient Greek conceptions or of biblical or Augustinian conceptions *or* of Hegelian or Marxist conceptions – Hannah Arendt commits a palpable blunder regarding the relationship between physical nature and human history. Her error is a touching one, because it is humanely motivated by a dread of the easy tolerance of a distinctly modern notion of history that Nazi totalitarianism could exploit, and indeed did exploit, to the shame of the whole world. She rightly sees that it is technology that bridges the difference between science and history. Understandably, she worries the conceptual marriage between a view of history in which the meaning of events is nothing but the instant purposes of human agents (a theory she assigns to Marxian *praxis*, which already prepares the ground for totalitarian opportunism) and an emergent technology (apotheosized in nuclear energy) that ushers in an age in which new but fundamental, hitherto unfamiliar, powers of nature are released, triggered, activated with utterly unforeseen consequences. In worrying that, she somehow supposes that the difference between physics and history is erased and that this has been accomplished by the new technology – which, therefore, poses for us the henceforth permanent threat of the true totalitarian insight: that every purpose, no matter how mad, can now be realized; and that history can do no more than stand by and acquiesce.

Of course, in saying this, it needs to be said at once that the use of Arendt's argument is intended here as a deliberate, somewhat tendentious economy for isolating certain conceptual questions regarding history and technology. It would be manifestly unfair to suppose that Arendt's full theory of history (as well as the full import of her reflections on technology) are adequately represented in the brief paper here exploited. Nevertheless, with that proviso, the mistake claimed is not a negligible one, and those who would counter the charge at stake are themselves duty bound to meet its terms. If anything may be drawn from the example of Arendt's ingenious account, it is surely that the

Paul T. Durbin (ed.), Technology and Contemporary Life, 217–236.
© 1988 *by D. Reidel Publishing Company.*

nature of history is difficult to formulate even for the most fastidious. It remains true, nevertheless, that the relationship between history, science, and technology here developed is substantially in agreement with Arendt's emphasis on the ideological dimension of science and, in particular, on the ideological cast of technology and its bearing on both the theory of science and the theory of history. Admitting the latter thesis, however, does not commit us to Arendt's specific account of the relationship between science and history.

Here, then, is the heart of her complaint – and claim: "If, therefore, by starting natural processes [she says], we have begun to act *into* nature, we have manifestly begun to carry our own unpredictability into that realm which we used to think of as ruled by inexorable laws."[1] Modern history, Arendt insists, opposes the Greek and Christian views (themselves opposed to one another) because, "For us . . . history stands and falls on the assumption that the process in its very secularity tells a story of its own and that, strictly speaking, repetitions cannot occur."[2] Arendt takes it as an irony of history that contemporary technology, now that it permits us to "act *into* nature," now that "for the first time [we] have taken nature into the human world as such and *obliterated* the defensive boundaries between natural elements and the human artifice by which all previous civilizations were hedged in,"[3] causes the idiographic, unpredictable features of history to infect the very structure of physical nature itself; and, because of that and because of the power of the entailed technology, we are (now said to be) permanently hostage to totalitarian drift: we are pressed into a mass-society and (conflating historical purpose and the meaning of history) we "have lost the world once common to all [mankind]."[4]

There is no way to assure Arendt or ourselves that her fear is not justified – even more and more justified with time. But her reasoning is wrong. She is wrong about the distinction between physics and history and therefore about the bridge role of technology. She has collapsed the conceptual issue and the heartfelt political fear of a technology increasingly out of control. In fact, her genuine dread of the future – no one who reads her can miss it – is at least in part an artifact of her mistake: so that, correcting *that*, we may possibly lighten *it*, at least to the extent of focusing our fears in a rational way.

Her strategy obviously pivots on a misreading of the import of Werner Heisenberg's uncertainty principle – to which she repeatedly returns[5] – though it is of course possible that Heisenberg himself failed to fix with suitable precision the message of his own discovery. Her way of reading the point of Heisenberg's discussion of the uncertainty principle leads her to conclude that we are "led into a situation in the sciences themselves in which man has lost the very objectivity of the natural world, so that man in his hunt for 'objective reality' suddenly discovered that he always 'confronts himself alone.'"[6] This (so far not unreasonable interpretation) she somehow construes as signifying the "obliteration" of the distinction between nature and history; whereas what it actually signifies is the untenability (clear in any case from the contemporary criticism of the correspondence theory of truth and its allied forms of essentialism, foundationalism, naive realism and the like) of construing the realism of science in terms that ignore the indissoluble symbiosis of the realist and idealist aspects of inquiry – now, increasingly and correctly historicized, with lamentably uncertain implications.[7]

Whatever may be drawn from that confusion and the altered picture of science, it does *not*, of itself, support the merging of the notions of history and nature or the conflating of that with our justified fears regarding the new potencies of an historically emergent technology. Heisenberg's principle does *not* endorse the blurring of nature and history: it signifies (at the present time) the impossibility of pursuing physics in any way not indissolubly encumbered by our historicized conceptions of physical nature; *and* by our appreciation of the global truth, that our very inquiry into nature affects the nature we would know. The latter point is true because of physics, not because of history, although our impulse to intrude and our grasp of what we are doing are due to history and not merely to physics. Combining these two considerations, we may well reflect on what, regarding the canonical picture of the laws of nature, might have led Arendt – or anyone similarly impressed or discouraged by the power of modern technology – to conflate history and nature. We might for instance be attracted to a distinctly different theory of the nomologicality of physical nature; and with that, to a firmer and more ramified theory of technology and of the problems of responsibility that would entail. But, for the moment, we may simply affirm that Arendt's maneuver is hardly due to her Heideggerian influences, also obviously occupied with the import of technology.[8]

In fact, Heidegger, who was also familiar with Heisenberg, never confuses the difference between history and nature, even though it is fundamental to his phenomenology that the disclosure of the structured world of entities other than *Dasein* (in effect, the domain of science) is symbiotically dependent on the reflexive and historicized inquiries of *Dasein* itself. The radical distinction between the "ontical" nature of things more or less conformable to "natural kinds" and of things that can be identified only "ontologically" (in terms of an existential concern for *Sein*) – what has been ingeniously contrasted as *Was-sein* and *Das-sein*[9] – signifies a disjunction between what lacks a history but is subject to natural law and what historicizes itself as well as the search for nature but cannot be suitably captured by subsumption under some natural kind. There may be a confusion between nature and history, in the sense in which *Dasein* (on Heidegger's view) may characterize itself as a "mere thing," adopting the categories of nature in some so-called "existentiell" reflection; but the union of nature and history obtains only at that "primordial" (*ursprünglich*), unanalyzable point at which *Sein* is first "disclosed" in terms of an array of what appears as *Da-sein* and what is utterly unlike *Dasein*,[10] that is, at a point logically prior to any and all such distinctions. Heidegger, therefore, hardly encourages Arendt's reading of Heisenberg's principle. And Heisenberg, certainly in his early interpretation of the uncertainty principle (that is, on the so-called Copenhagen interpretation), took it that the new physics reflected inherent limitations (apparently insurmountable) in human *knowledge*: whatever the real world might be supposed to be, every effort to capture it cognitively was confronted (so it was claimed) by an objective barrier that could not be overcome.[11] Whatever else may be made of this claim, it does *not* entail a confusion between history and physics: on the contrary, it retreats, in effect, to the history of inquiry in order not to confuse the limits of knowledge with the structures of the real world.

Part of the trouble undoubtedly lies with Arendt's insistence on the idiom of thinking of physical nature as ruled by "inexorable laws." It is curious that the very same phrase occurs in Karl Popper's well-known attack on historicism – though it occurs there in scare quotes and in reference to the would-be laws of economics rather than of physics, obviously because of an intended reference to Marx.[12] Popper's purpose is to concede, at least in principle, that the *laws* of such disciplines as economics and sociology have the same status as the laws of physics, as against the would-be laws of history (which Popper regards as a contradiction in terms). Popper himself tends to construe historicism (rather heterodoxly) as "an approach to the social sciences which assumes that *historical prediction* is their principal aim, and [more orthodoxly] which assumes that this aim is attainable by discovering the 'rhythms' or the 'pattern,' the 'laws' or the 'trends' that underlie the evolution of history."[13] He has of course often discussed the central puzzles regarding the laws of nature; but rather compendiously, in assessing the pertinent theories of Auguste Comte and John Stuart Mill – in particular, applauding their devotion to the ideals of science within the scope of the human or social sciences, while at the same time repudiating their commitment to what Mill has called "the law of progress," in effect the lawlike regularities of history by means of which future events may be predicted, by analogy (by a bad analogy) with predicting successive numbers in an arithmetic series[14] – Popper favorably mentions "Comte's emphasis on laws and scientific prediction, . . . his criticism of an essentialist theory of causality; and . . . his and Mill's doctrine of the unity of scientific method. Yet their doctrine of historical laws of succession [he adds] is, I believe, little better than a collection of misapplied metaphors."[15]

Between Arendt and Popper, then, we span the principal ways of failing to grasp the conceptual role of technology and its bearing on disputes about public responsibility, essentially because of their peculiar (but different) ways of collapsing the distinction between nature and history – however unlikely that may seem to be. They approach matters from opposite ends of the tunnel: Arendt sees the historical thrust of technology altering the essential structure of physical nature, hence also the logical properties of physical laws; and Popper sees the unity of scientific method ranging inexorably over physical nature, hence as setting essential constraints on the rational possibilities of technology. Arendt overextends the powers of human intervention, because of her fears of the permanent threat of totalitarianism; and Popper sets excessive constraints on those powers – ironically, in the very name of preserving human freedom[16] – by rejecting the conceptual possibility of historicist laws. Both misunderstand the nature of scientific laws and, because of that, the prospects of technology.

II

Popper formulates his thesis about the nature of laws in the strongest way possible – in accord, that is, with his own falsifiability thesis (hence, in the strongest way opposed to an inductivist [or Humean] reading of

physical law): rejecting the very notion of laws of history and contrasting historical *trends* and physical *laws* in a completely disjunctive way,
Popper holds that "A statement asserting the existence of a trend is
existential, not universal. (A universal law, on the other hand, does not
assert existence; on the contrary . . . it asserts the impossibility of
something or other [that is, of anything inconsistent with the universalized nomic connection].) And a statement asserting the existence of a
trend at a certain time and place would be a singular historical statement, not a universal law."[17] A would-be physical law, then, is universal, deterministic, and inviolate; for it makes no existential reference
(though it will of course have existential application); and it *"asserts"*
the universal impossibility of a physical event contrary to its putative
nomic connection (hence, is falsifiable and *not*, in Popper's account,
"metaphysical").[18] Popper also denies that we are able to discover such
laws (against the inductivist) *and*, at the very same time (though quite
incoherently, always in accord with his falsifiability thesis), he claims
that we may count on nomic verisimilitude, that is, on the progressive,
asymptotic approximation to such laws.[19] This captures the sense in
which Popper subscribes to the correspondence theory at the same time
he denies that we have any direct (epistemic) access to assessing such
correspondence.[20] A so-called anti-realist (of Michael Dummett's
sort[21]) would of course have denied Popper the right to hold such a view
of laws if he denied to himself (as he does) any first-order ability to
"decide" the truth value of the pertinent approximative claims. But this
issue invites a merely technical skirmish that need not be permitted to
deflect us from the principal contest.

The essential point is that both Arendt and Popper seriously miscon-
strue the nature of laws in science – and, because of that, the nature of
human history and human technology. Arendt finds the laws of physical
nature so permeable by the intrusions of human technology that their
logical properties can actually be altered, however inadvertently, by
intrusions of human purposiveness; and Popper finds the laws of physical nature so fixed and homogeneous that the causal processes of human
history and technology are explicable in terms of precisely the same
method that holds in the physical sciences. This is the reason that, in
spite of his well-known opposition to the inductivism of the positivists
(and of Millian social scientists), and in accord with his insistence that
"concrete social situations are in general less complicated than concrete
physical situations," Popper asserts that "all theoretical or generalizing

sciences make use of the same method, whether they are natural sciences or social sciences."[22] The truth is that he adheres, just as tenaciously as ever Carnap (or Hempel) did, to what has come to be called the hypothetico-deductive model of explanation: "I suggest [he says] that to give a causal explanation of a certain *specific event* means deducing a statement describing this event from two kinds of premises: from some *universal laws*, and from some singular or specific statements which we may call the *specific initial conditions*."[23]

It is perhaps because of this adherence to the unity of method that Popper introduces his completely disastrous contrast between "historical trends" and "scientific laws." Here is his charge: ". . . Mill and his fellow historicists *overlook the dependence of trends on initial conditions*. They operate with trends as if they were unconditional, like laws. Their confusion of laws with trends makes them believe in trends which are unconditional (and therefore general); or, as we may say, in 'absolute trends'; for example, in a general historical tendency towards progress – 'a tendency towards a better and happier state'. . . . This, we may say, is the central mistake of historicism. Its 'laws of development' turn out to be absolute trends which, like laws, do not depend on initial conditions, and which carry us irresistibly in a certain direction into the future. They are the basis of unconditional *prophecies*, as opposed to conditional scientific *predictions*."[24]

What Popper fails to grasp is that *if* there are formulable laws of nature, then such laws are inextricably encumbered by initial existential or indexical conditions of reference, in precisely the same way trends are; or, *if* singular reference can be eliminated in principle with respect to natural laws, then so can they with respect to trends and for the very same reasons. The upshot is: either laws are *not* deterministic universals that (by a falsifiability interpretation) hold only if it is physically or naturally impossible that anything obtain contrary to what they "assert";[25] or the line of demarcation between laws and trends cannot be maintained, and laws of history – historicist laws – are not precluded (at least) for the reasons Popper alleges. To recover a more balanced picture of laws and trends is to recover a more balanced picture of technology; and thus, almost casually, to correct the logical point of Arendt's fears, though not of course to deny the social function of such fears. Also, the correction would at one stroke utterly disqualify Popper's conception of science, since *if* the laws of nature are not strictly universal or if (what comes to the same thing) the laws of nature are existentially encumbered in the way trends are, then the falsifiability interpretation of scientific method *cannot be epistemically defended in principle*, whether or not inductivism can. The status of Popper's theory of science is, of course, not our direct concern here; nor is the defense of inductivism – which as it happens (as in Hempel's account of the deductive-nomological model of explanation) suffers from precisely the same logical difficulties that plague Popper's conception of natural laws, despite important methodological differences between the two accounts. But *if* laws are existentially encumbered, then the falsification of unencumbered universals *is not* tantamount to the falsification of statements of encumbered regularities. On Popper's view, *this would entail, scandalously, that scientific laws were thoroughly metaphysical*.

The required demonstration (*contra* Popper) is straightforward enough; it involves two steps: one, to show that historical or existential encumbrance is intrinsic to science as a human undertaking; the other, to show that the elimination of singular reference cannot be disjunctively applicable to laws and trends. The upshot would be a radically altered theory of the constraints on human history and human technology.

Consider, for instance, that the analytic tradition from, say, Bertrand Russell to W. V. O. Quine believes itself to have managed – in a purely formal way – to eliminate singular reference in favor of uniquely satisfied general predicates. That is, on a Leibnizian view of the identity of indiscernibles, Quine has proposed that we treat proper names (in effect, the intended function of singular referring or designative expressions) as general predicates – on the model of "pegasizing."[26] The difficulty this occasions is a double one: for one thing, we cannot construe such predicates as interpreted in any cognitively pertinent sense, in spite of the fact that the referring expressions they replace do succeed, in the context of actual use, in functioning referentially; and for another, there is absolutely no basis for supposing that, within the real-time constraints of human discourse, we could ever provide an interpretation of such predicates that either captured their unique intent or effectively approximated it. But *if* it were argued that this *could* be done "in principle," then, *contra* Popper, the difference between regularities collected as "laws" and as "trends" could not be supposed to be distinct on that score.

Of course, the Quinean project presupposes that the Leibnizian strategy *can* be managed in transparently extensionalist terms; but

Quine supplies no cognitively pertinent reasons for believing this feasible, even "in principle," or for believing that complexities intrinsic to the use of natural languages would ever yield completely (or "adequately") to an extensionalist paraphrase. But if something like a Quinean program is impossible, or impossible in real-time terms, then it cannot be maintained that the contingent, historicized, existentially motivated inquiries of human agents – intended to lead in a cognitively responsible way to the formulation of general laws – *do* reliably lead to laws lacking indexical or singular referring clauses. Hempel, for example, is very emphatic about the methodological unity of science; the preeminent role of universal laws both in explanation and prediction; the complete neutrality in these respects of "specifically historical laws" or sociological laws or the like; *and* the essential point – that explanation in accord with the deductive-nomological model requires that the relevant descriptive terms "stand for kinds or properties of events, not for what is sometimes called individual events. For the object of description and explanation in every branch of empirical science is always the occurrence of an event of a certain kind. . . ."[27] In fact, Hempel (rightly) chides the philosopher of history, Maurice Mandelbaum, for failing to grasp the force of the requirement[28] – a mistake which is all the more odd as Mandelbaum professes to subscribe to the so-called "covering law" model, that is, less ambiguously, to the unity of science model.[29] In any event, on a Humean or inductivist view (which Hempel accepts), would-be laws of nature are in some sense "summative" or idealized from empirical summations *of* observed particular events: in that sense, the elimination of indexical or singular referring clauses poses an *epistemic* problem that cannot be resolved solely by attention to the allegedly deductive nature of scientific explanation or to the "independent" logic of scientific laws.

We have already drawn a similar finding from Popper's anti-inductivist or falsificationist theory of laws. Popper differs from Hempel in a somewhat verbal way, in that *he* is willing to speak of the "historical" aspect of both genuinely historical explanation *and* physics, wherever the *explanandum* is a singular event.[30] But, like Hempel, Popper makes it quite clear that the boundary conditions under which a *law* obtains (as opposed to a trend) "are *not* singular, and refer to a certain *kind* of situation."[31] It is true that, on Popper's view of laws, it is easier to construe a would-be law as altogether lacking singular or indexical clauses, simply because (on that view) candidate laws "assert" what is physically or naturally impossible. But it is also Popper's view that we are never in a cognitive position to affirm any particular law securely (only to falsify it) – and Popper's approximative view of verisimilitude affords absolutely no cognitive grounds for treating any formulation as an actual law: it is more nearly a heuristic target for organizing the provisional achievements of our science, which *do* accommodate reference to singular (falsifying) events. In any case, Popper's notion of verisimilitude is, on Popper's own grounds, a "metaphysical" doctrine (as being not falsifiable); and Popper himself is a confirmed indeterminist and committed to the inherent "incompleteness" (that is, non-closure) of the physical world (with respect to psychological and cultural influences).[32]

III

These considerations strongly suggest that the unity of science conception of laws – generically construed, in a sense neutral to such disputes as that between inductivists and anti-inductivists – is simply an undefended dogma. J. J. C. Smart concedes that there are no biological laws, on the lines Hempel favors;[33] the truth is that there are no laws at all, in the canonical sense. In what is perhaps the most recent, sustained account of scientific laws within the unity tradition, D. M. Armstrong affirms:

For it to be a law that an *F* is a *G*, it must be *necessary* that an *F* is a *G*, in some sense of 'necessary'. But what is the basis in reality, the truth-maker, the ontological ground, of such necessity? I suggest that it can only be found in *what it is to be an F and what it is to be a G*. . . . We need also to construe a law as something more than a mere collection of necessitations each holding in the individual case. How is this to be done? I do not see how it can be done unless it is agreed that there is something identical in each *F* which makes it an *F*, and something identical in each *G* which makes it a *G*. . . . But this is to say that the necessitation involved in a law of nature is a relation between universals.[34]

But this only shows (at least) what upholding the unity thesis entails (or related views, for instance, the need to appeal to laws to "explain regularities"[35]). It does not show the least reason, *epistemically*, why we should construe laws thus. It introduces two notoriously dubious doctrines: that of the necessary relational connections between properties nomologically linked,[36] and that of a strong identity of real properties or universals multiply instantiated.[37]

But it is hardly difficult to hold that selected regularities may be construed as laws (provisionally at least) because of their strategic connection with explanatory theories (perhaps disputably or variably, as phenomenological or explanatory laws[38]) and because of the state of our technological powers (invention and experiment, particularly[39]), in spite of the fact that: (i) they are not strict universals, (ii) they are epistemically (not merely contingently) linked to the existential circumstances of inquiry, (iii) they are not adequately justified on either inductivist or falsificationist grounds, (iv) they do not commit us to nomic necessity in any clear way, and (v) they throw into extreme doubt the very model of deductive-hypothetical explanation (without reference, incidentally, to probabilistic considerations). Furthermore, it is always open to us, if we wish, to idealize general causal regularities (that is, what purport to be such regularities) in the canonical form favored for deductive-nomological explanation; all we need concede is that such an idealization is heuristically and epistemically dependent on some more fundamental human activity – molar, praxical, historicized, technological, horizon-bound, prejudiced, problem-centered, paradigm-oriented, interest-driven, or constrained in some similar way – in virtue of which the theoretically privileged status of nomic universals is simply repudiated. *In effect, to construe human inquiry and science as technologically or praxically encumbered is to give up, except heuristically* (or, in Popper's idiom, in some "metaphysically" verisimilitudinous sense), *the formulation of nomic universals.*

There are, however, more heterodox possibilities a-lurking. Consider that, on Hume's view, there is no rational justification for causal necessity or nomologicality: it is certainly strange that the inductivist version of the deductive-nomological model of explanation should not have faced up to the *epistemic* problems of its own methodology. Consider that, on Popper's view, nomic universals, like scientific theories in general, are "human inventions – nets designed by us to catch the world," never to be "mistaken for a complete representation of the real world in all its aspects; not even if they are highly successful; not even if they appear to yield excellent approximations to reality"[40]: it is certainly strange that, admitting the "metaphysical" (that is, the empirically untestable) nature of scientific determinism – "we must [Popper says] be metaphysical *indeterminists*, [though] methodologically we should still search for deterministic or causal laws"[41] – Popper should have insisted on a disjunctive demarcation between scientific laws and historical trends *that could only have depended on untestable grounds*, or should have insisted, just as exclusively as the inductivists, on the deductive-nomological model of explanation, *the rationality of which depends entirely on metaphysical grounds*. Consider that D. M. Armstrong effectively admits that, although "[he] should like to believe that all causation is governed by law," he does "not see how to exclude the logical possibility of causation without law," the possibility that causality does not entail nomologicality[42]: it is strange that he proceeds to develop a view, favorable to the thesis of nomic universals (in fact, of nomic universals instantiating "nomic necessitation," itself construed as a "primitive" relation) – to the effect that "If *a*'s becoming *F* causes *b* to become *G*, then there is an *F–G* law only if the first event causes the second event *in virtue of the universals F and G*" – *without attending at all to the epistemic conditions under which causal and lawlike relations may be supported, and without exploring the theory of causation itself.*[43] And consider, finally, that Donald Davidson distinguishes between the extensional contexts of causality (of events functioning causally) and of the intensional contexts of causal explanation (of the function of causal statements, under suitable descriptions of causes), acknowledges (with C. J. Ducasse) that "singular causal statements entail no law and that we can know them to be true without knowing any relevant law," *and* adds (with Hume, as he reads Hume) that singular causal statements *do* "entail there is a [covering] law":[44] it is strange that Davidson never addresses the question of why (or even in what sense it is true that) *if* covering laws are required for causal *explanation* and if explanations are good ones "only if" they accord with what actually causes what,[45] *causality entails nomologicality*.

IV

It is difficult to avoid the finding that the various versions of the unity of science view of causal laws are distinctly prejudiced and almost completely without foundation – which of course is neither to deny that science has actually progressed in formulating well-confirmed causal laws nor to deny that the idealization of such laws along universal and necessaritarian lines, laws lacking existential or indexical clauses, may even be essential to the projects of scientific explanation. But there is no contradiction here: the assigned features may have an important heuristic and systematic function. What strikes the mind is that, for all the difference between the inductivist and anti-inductivist, there is a strong convergence between them – toward universality, toward the absence of existential encumbrance, toward nomic necessitation: in effect, the inductivist (somewhat against Hume) seems to find epistemic grounds for his position;[46] and the anti-inductivist (ironically, also in accord with Hume, though quite contrary to his inductivism) holds that there are no epistemic grounds for affirming laws of the requisite sort. Both, perhaps, are guided by a strong conviction that causal laws entail distinctive counterfactuals.[47]

There can be no question that the Western tradition has favored versions of *this* (unity of science) conception of causality. But there is another source of the notion, a source of an utterly different kind, extraordinarily natural in *any* terms that emphasize the molar interests of human persons, the historical and praxical contingencies of inquiry and human intervention in nature; it yields a notion that is certainly not inevitably committed to any necessitarian doctrine (to any doctrine of "inexorable laws") – like that, as we saw, Hannah Arendt embraces or Popper permits to be formulated "metaphysically" or solely for purposes of being falsified.

Very simply put, the alternative holds that it is an idea inseparable from the very concept of a person *that a person be an agent, a causally effective agent.*[48] To understand what it is to be a person is to understand how, in what a person does – acting rationally, in accord with plans and preferences and the like, intentionally pursuing objectives, successfully surviving as a *consequence* of such intervention – an agent causally brings about certain states of affairs as a direct result of having acted. *Human history and human technology are premised on that capacity*, however much they cannot be exclusively analyzed in terms of mere agency, and however problematic the conceptual connection between the two notions may be. Only if persons could be ontologically eliminated or reduced to the processes of the inanimate physical world or to those of biology below the level of incipient cognition and rationality,[49] could we expect to eliminate the causal notion that (we are claiming) is intrinsic to the very notion of a person.

We cannot here pursue the full import of this thesis. But we need not, we need only consider the most salient relevancies. Certainly, the agency conception (as we may term it) makes utterly problematic the idea that causality entails nomologicality – even if, following Ducasse, Davidson, Anscombe, Armstrong and others, it is conceded that we may know singular causal statements to be true without knowing any relevant law. The very concept of a person commits us to admitting any of a very large number of familiar instances (and of kinds of instances – *not* nomically identified) in which human agents effectively produce intended consequences: hunting and fishing, eating and cooking, making tools and the like. We cannot say (certainly we cannot say on such grounds alone, possibly even in principle) that human agency precludes nomologicality in the strong sense intended in the unity of science view. But the agency thesis shows that causality and nomologicality are independent notions; hence, that causality and nomic necessitation are independent notions. *If*, further, it proves impossible *always* to be able to identify and reidentify agency extensionally (because, say, what an agent does *in acting* deliberately or the like – *and, effectively, because of so acting* – can only be identified in intensionally complex ways), then, contrary to Davidson at least, causal contexts may not behave in an invariably extensional way (where, by "extensional," one means that *we* should be able to pick out items extensionally, not that we merely affirm that everything is identical with itself).[50]

Again, it will usually be the case that what one perceives or reflexively understands to be causally efficacious in the agency sense (hitting a baseball, for instance), or what one perceives to be an event possibly due in a causal way to human agency (perceiving *that* that window *has been broken*, for instance), is first identified in intensionally pointed terms. The reason is that ascriptions of human action presuppose some conceptual dependency on understanding what human rationality entails – *some* species-wide and culturally differentiated order of wants, desires, perceptions, beliefs, intentions, skills, and the like, linked holistically in however minimal or loose a manner, but individuated as such primarily in whatever intensional terms may be reflexively (and also critically) favored by the agents in question.[51] Whether, therefore, we can, in all such instances, rephrase what has been done (by way of human action) so that a minimally adequate set of "basic actions" or "primitive actions," identified in purely physicalist and extensionalist ways, can replace our admittedly colorful but intentional argot remains a distinct puzzle.[52]

Furthermore, *if* causality and nomologicality are separated, *if* agency cannot be invariably or even for the most part made to yield to extensionally perspicuous causal replacements, then the explanation of the phenomena of the human world will have to be formulated in ways hospitable to models utterly different from that of the deductive-nomological model. So we can see the completely subversive import of admitting the irreducibility of the concept of a human person or of acknowledging, with that, the logical peculiarities of the notion of agency. But in conceding that, we need hardly confuse (with Arendt) the distinction between history and physics; nor need we radically disjoin (with Popper) laws and trends.

The distinctive feature of human history – in the double sense of a series of actual events and of a narrative of such events – is its intensional structure: the primary events of history are surely human actions and what they produce. But to say that is already to identify the essential difficulty of speaking of causal laws in the human sciences (however diverse those sciences may be from one another). We have already conceded the sense in which physical laws need not be universal, unencumbered by existential or referential clauses, or necessitarian; and by implication, we have conceded that, *if* discourse about persons is not reducible or eliminable in physicalist terms, then the very domain of physical events cannot be completely closed homonomically.[53]

Nevertheless, the working assumption of the physical sciences holds (not unreasonably) that the regularities of all would-be covering laws can be empirically provided *in accord with strongly or strictly extensionalist procedures*. So seen, the distinction of the human sciences is precisely that the counterpart assumption is doubtful or at least open to fundamental dispute. Popper's simplistic contrast between laws and trends obscures the issue: it is not a question of indexical constraints on trends; it is a question rather of the difficulty of formulating *either* laws or trends if the regularities at stake cannot be regimented in the extensionalist way. This is why Popper's side remark that social phenomena are less complex than physical phenomena is so revealing: Popper obviously has not quite grasped the peculiar puzzle of would-be sociological laws or would-be historical trends. One has only to think for instance of attempts to generalize about the causal dynamics of political revolution,[54] or of the significance of post-Keynesian economic habits in the United States, to see the point: the laws and trends of the human sciences are *conceptually inseparable from an historically contingent interpretive effort to identify what is intensionally common to a set of culturally pertinent phenomena, regarding which the further intensionally similar uniformities of selected instances of human agency may be claimed to support would-be laws or trends.* The human sciences are marked by the absence of any straightforwardly extensionalist procedures *for* fixing

the events we claim behave in lawlike ways; and the procedures that we
do have, for cataloguing phenomena relevantly to this end, are *themselves* notably affected in divergent ways by shifts of history – are
themselves phenomena of the same sort they help to explain and are
open therefore to similar causal explanation. There is no exact counterpart in the physical sciences, in spite of the fact that, *as* human undertakings, all science is subject to historical contingency. This is why talk of
prediction in the human sciences – while not inadmissible – is essentially
complicated by an interpretive intrusion of an historically biassed but
ineliminable sort[55] logically different, therefore, from what is ideally
meant by prediction in the physical sciences.

Technology, then, bridges (in the double sense the term "history"
signifies) the difference between history and physical nature and between history and physical science. It provides the conceptual matrix
within which human agency is, under historical conditions, reflexively
guided by what it takes to be the lawlike regularities of nature and of its
own forms of life. These, the sciences, tend to bifurcate: the physical
sciences appear to yield to an extensionalist methodology, however
encumbered historically by the inherent constraints of human inquiry;
the human sciences appear to be intensionally recalcitrant, in the sense
that (failing a pertinent reductionism) the admission of persons precludes a correspondingly extensionalist treatment of the lawlike regularities of human phenomena. This is why the unity of science
conception of laws is at least heuristically useful in the physical sciences,
apart from local quarrels between inductivists and anti-inductivists; and
this is why the agency conception is ineliminable in the human sciences,
despite methodological consequences we have barely begun to sketch.
Technology, then, is (or addresses) the context in which *historical
agency pursues nomological regularities within the contingent boundaries
of a praxis that survives in the continuing short run*. It makes human
responsibility intelligible, therefore, in terms of human intrusions into
nature, in terms of the largest seeming regularities of human existence
itself, in terms of historical reflection upon and revision of human
objectives, *and* in terms of the conceptual coherence and linkage holding among these distinct undertakings.

Arendt fears that we "have lost the world once common" to all
mankind, because of our technological intrusion into nature and because, in intruding, we have historicized the laws of nature itself. And

Popper overrides the possibility, and thus the need, for sorting distinctive laws within the human sciences; and in doing that, bifurcates and thus renders more arbitrary, or at least more alien, the link between the explanation and would-be rational appraisal of human behavior itself.[56] But the model of human rationality, holistic and intensionalized, historically variable and subject to change within reflexive speculations regarding species-wide concerns, aptitudes, dispositions, needs and the like, constrains all our undertakings – method in the extensionalized sciences, objectivity and scope in the human sciences, survival, responsible and informed agency – in different but systematically connected ways. Agency itself, attenuated increasingly in terms of apt analogies within the animal world and even inanimate nature (think for instance of the "action" of the surf on the shore) is increasingly extensionalized and universalized in accord with the idealization of the unity model: so a smooth declension is possible from the one model to the other, *but only in one direction* if the argument here advanced be admitted. That asymmetry is what keeps the notion of technology itself from collapsing into a merely applied science: where, that is, "science" conforms with the constraints of the unity model, "causality" is relieved of anthropomorphic links with human agency, and "applied" is a term externally focused on selections, however arbitrary or independently guided, by which the achievements of science (otherwise "closed") are made to serve particular human objectives.

Temple University

NOTES

[1] Hannah Arendt, 'The Concept of History: Ancient and Modern', in *Between Past and Future* (enlarged ed.; Harmondsworth: Penguin Books, 1968), p. 61.

[2] *Ibid.*, p. 67.

[3] *Ibid.*, p. 60; italics added.

[4] *Ibid.*, p. 90.

[5] Cf. *ibid.*, pp. 48–49, 86–87; also, 'The Conquest of Space and the Stature of Man', *ibid.*, pp. 276–280.

[6] 'The Conquest of Space and the Stature of Man', p. 277; the cited material is from Werner Heisenberg, *The Physicist's Conception of Nature*, trans. Arnold J. Pomerans (London: Hutchinson, 1958), p. 24.

[7] I have pursued this question, in considerable depth, in *Pragmatism without Foundations* (Oxford: Basil Blackwell, 1986).

[8] See Martin Heidegger, *The Question concerning Technology and Other Essays*, trans. William Lovitt (New York: Harper and Row, 1977).

[9] Cf. the suggestion in the note, supplied by David Farrell Krell in his edition of Martin Heidegger, *Basic Writings* (New York: Harper and Row, 1977), to the translation of the Introduction of *Sein und Zeit*, p. 48.

[10] Martin Heidegger, *Being and Time*, trans. (from 7th ed.) John Macquarrie and Edward Robinson (New York: Harper and Row, 1962), § 45.

[11] See for instance Karl R. Popper, *Quantum Theory and the Schism in Physics* (*Postscript to the Logic of Scientific Discovery*, vol. III), ed. W. W. Bartley, III (Totowa, N. J.: Rowman and Littlefield, 1982), Preface 1982, Introductory Comments.

[12] Karl R. Popper, *The Poverty of Historicism* (3rd. ed.; New York: Harper and Row, 1961), p. 7.

[13] *Ibid.*, p. 3.

[14] John Stuart Mill, *A System of Logic* (8th ed.; New York: Harper, 1874), Bk. VI, Ch. X, sec. 3; cited by Popper.

[15] *Op. cit.*, p. 119.

[16] See Karl R. Popper, *The Open Society and Its Enemies* (Princeton: Princeton University Press, 1956).

[17] *Ibid.*, p. 115.

[18] Cf. Karl R. Popper, *The Logic of Scientific Discovery* (New York: Basic Books, 1959), §§ 4–6, 85.

[19] See Karl R. Popper, 'Two Faces of Common Sense', in *Objective Knowledge* (Oxford: Clarendon, 1972).

[20] See Karl R. Popper, 'Philosophical Comments on Tarski's Theory of Truth', in *Objective Knowledge*.

[21] See Michael Dummett, *Truth and Other Enigmas* (Cambridge: Harvard University Press, 1978).

[22] *The Poverty of Historicism*, §29, especially pp. 140, 130.

[23] *Ibid.*, p. 122. See also, Carl G. Hempel, 'The Function of General Laws in History', in *Aspects of Scientific Explanation* (New York: Free Press, 1965).

[24] *Ibid.*, p. 128.

[25] Cf. *ibid.*, p. 146.

[26] W. V. Quine, *Word and Object* (Cambridge: MIT Press, 1960), §37.

[27] *Op. cit.*, p. 233.

[28] *Ibid.*, p. 241, note 7.

[29] Cf. Maurice Mandelbaum, *The Problem of Historical Knowledge* (New York: Liveright, 1938), pp. 13–14 (cited by Hempel); and 'The Problem of "Covering Laws"', *History and Theory*, I (1961), which unaccountably compounds the error.

[30] *The Poverty of Historicism*, pp. 144–145.

[31] *Ibid.*, p. 125.

[32] Cf. *Quantum Theory and the Schism in Physics*, chap. 4; and Karl R. Popper, *The Open Universe* (*Postscript to the Logic of Scientific Discovery*, vol. II), ed. W. W. Bartley, III (Totowa, N. J.: Rowman and Littlefield, 1982), particularly Addendum I; and 'Two Faces of Common Sense'.

[33] J. J. C. Smart, *Philosophy and Scientific Realism* (London: Routledge and Kegan Paul, 1963), chap. 3; cf. also Ernst Mayr, *The Growth of Biological Thought* (Cambridge: Harvard University Press, 1982), chap. 2.

[34] D. M. Armstrong, *What Is a Law of Nature?* (Cambridge: Cambridge University Press, 1983), pp. 77–78.

[35] *Ibid.*, p. 41.

[36] *Ibid.*, p. 39.

[37] Cf. D. M. Armstrong, *A Theory of Universals*, vols. 1–2 (Cambridge: Cambridge University Press, 1978); also, Joseph Margolis, 'Berkeley and Others on the Theory of Universals', in Colin Turbayne, ed., *Berkeley: Critical Essays* (Minnesota: University of Minnesota Press, 1982).

[38] See Nancy Cartwright, *Why the Laws of Physics Lie* (Oxford: Clarendon, 1983).

[39] See Ian Hacking, *Representing and Intervening* (Cambridge: Cambridge University Press, 1983).

[40] *The Open Universe*, pp. 42–43.

[41] *Ibid.*, p. 149.

[42] *Op. cit.*, p. 95; Cf. also D. M. Armstrong, *A Theory of Universals*, vol. 2, p. 149, cited by Armstrong himself.

[43] *What is a Law of Nature?*, pp. 95 (including note 7), and 88.

[44] Donald Davidson, 'Causal Relations', in *Essays on Actions and Events* (Oxford: Clarendon, 1980), p. 160. Cf. also, C. J. Ducasse, 'Critique of Hume's Conception of Causality', in *Causation and the Types of Necessity* (Seattle: University of Washington Press, 1924), cited by Davidson; and G. E. M. Anscombe, *Causality and Determination* (Cambridge: Cambridge University Press, 1971).

[45] See for example Davidson's sympathetic but careful adjustment of Hempel's views, in 'Hempel on Explaining Action', *op. cit.*, particularly pp. 262, 264, 265.

[46] See for instance Hans Reichenbach, *Laws, Modalities, and Counterfactuals* (Berkeley: University of California Press, 1976).

[47] This cetainly is emphasized by Armstrong, *What is a Law of Nature?*, p. 168. The issue has of course exercised many: for instance Carl G. Hempel, 'Studies in the Logic of Confirmation', in *Aspects of Scientific Explanation*; and Nelson Goodman, *Fact, Fiction, and Forecast* (2nd ed.; Indianapolis: Bobbs-Merrill, 1965).

[48] I have given a brief sketch of the contrast between this notion of causality and the one we have been examining, in 'The Causal Explanation of Human Actions', in *Culture and Cultural Entities* (Dordrecht: D. Reidel, 1984).

[49] I have explored versions of such a strategy in a variety of places, most recently in *Philosophy of Psychology* (Englewood Cliffs, N. J.: Prentice-Hall, 1984); and in 'Eliminating Selves in the Psychological Sciences', in Polly Young-Eisendrath and James Hall, eds., *Self: Language, Experience, Construct* (New York: New York University Press, 1986).

[50] There is a very sympathetic sketch of this sort of holism in Davidson's 'Mental Events', in spite of the fact that it is Davidson's intention to obviate the need for reference to such models, within the bounds of science.

[51] Cf. Donald Davidson, 'Mental Events', in *Essays on Actions and Events*.

[52] This of course is the strategy favored by Davidson: cf. for instance, 'Agency', *Essays on Actions and Events*; and Arthur C. Danto, 'Basic Actions', *American Philosophical Quarterly*, II (1965). I have examined these views, in 'Action and Causality', *Culture and Cultural Entities*.

[53] Cf. Davidson, 'Mental Events'.

[54] See for instance Hannah Arendt, *On Revolution* (New York: Viking, 1965).

⁵⁵ The issue affects in a most decisive way the dispute raised and considered in Adolf Grünbaum, *The Foundations of Psychoanalysis* (Berkeley: University of California Press, 1984).
⁵⁶ The most extreme scenario, here, would insist on a completely laissez-faire view of constraints on human technology *beyond* the limitations of purely physical laws. Cf. Joseph Margolis, 'Three Conceptions of Technology: Satanic, Titanic, and Human', in Paul T. Durbin, ed., *Research in Philosophy and Technology*, vol. 7 (Greenwich, Conn.: JAI Press, 1984).

ANDRIES SARLEMIJN AND PETER A. KROES

TECHNOLOGICAL ANALOGIES AND THEIR LOGICAL NATURE

A superficial orientation in the literature is sufficient to recognize that analogies play a very important role in technological and scientific research.[1] We quickly reached this conclusion after starting with our study of this topic. Nevertheless we were surprised to see that there exists almost no methodological literature on the use of analogies in the technical sciences. So, it appeared that we had found a hole in the literature which could, we hoped, easily be filled.

However, appearances are deceptive. Starting from our different backgrounds, we analyzed several fields in which analogies play a role, but we reached disparate results. After some time, however, it became clear that this was due to the subject matter itself. This insight led to the conclusion (which is our most general conclusion) that indeed analogies play a very important role in the technical sciences but that they can be of a different nature or type. This confirmed Bochenski's observation that "the notion 'analogy' is itself analogous," even in a stronger sense than Bochenski intended in 1948.

In technological and scientific research fundamentally different notions of analogy occur, each with its own logical or set-theoretic definition and foundation. In the following, these notions of analogy will be analyzed. Sharp distinctions between various types of analogies will be introduced. On the basis thereof, the role of analogies in technological thinking will be discussed extensively. This leads to a better insight into the nature of technological thinking, in particular into the relation between theoretical ideas and technological developments.

1. ANALOGIES IN THE TECHNICAL SCIENCES

In this section we shall first sum up several tasks of technical scientists and describe in broad outline the notions of analogy corresponding with these tasks.[2] In the following sections, each notion of analogy will be treated separately and the differences between them will be pointed out.

(1) The technical scientist is expected to be able to apply fruitfully the *same* technological principle in *different* ways. In this wording we clearly

237

Paul T. Durbin (ed.), Technology and Contemporary Life, 237–255.
© 1988 *by D. Reidel Publishing Company.*

recognize the key idea behind any analogy: resemblance or similitude in different situations.

To illustrate this, we will have a brief look at the technical principles underlying the steam engine:

$$\text{heating} \rightarrow \text{expansion} \rightarrow \text{mechanical action}$$
$$\text{cooling} \rightarrow \text{contraction} \rightarrow \text{mechanical action}$$

Huygens, his pupil Papin, Savery, Newcomen and Watt, all used these same principles in different desigs and constructions. We will call the relation between these different applications ANAMORPHY (in order to contrast it with isomorphy).

Let us concentrate on what is specific to this technological-scientific context, namely the fact that more or less *general* technological principles are applied in different *concrete* situations in different ways. We see that the question arises whether the use of general technological principles in concrete situations can be given a foundation. Here we touch upon one of the central problems of inductive logic.

Inductive logic and anamorphy are also important in research which is more theoretically oriented, that is research which aims at the explanation of physical phenomena. Newton for instance tried to find the fundamental principles of optics starting from an analogy with acoustical phenomena. The same method was used by Huygens and Young. Maxwell searched, as we shall see later on in more detail, for a similarity between thermodynamics and mechanics starting from an analogy between both type of phenomena and a similarity in the relevant equations. So anamorphy (analogy with an inductive logical character) not only plays a distinctive role in the technologically oriented sciences, but also in the theoretical sciences.

In order to elucidate the specific character of anamorphy, the following should be noted. It is well known that conclusions based upon inductive reasoning always possess some uncertainty (this will be illustrated later on). In the case of anamorphy, this uncertainty manifests itself in a specific way. In teaching, for instance, the use of anamorphies has its positive and negative sides. The situation is often the following. An unfamiliar or new principle (usually called the recipient principle) is explained by using a principle (the donor principle) which is familiar to the students, in spite of the fact that not all the properties of donor and recipient principle are the same. The anamorphism can misguide the students when its inductive logical character is not clearly explained. In particular the differences between donor and recipient field should be underlined. Somewhere the analogy will go wrong. To the student this may seem rather abstract, but these theoretical exercises are necessary for acquiring well-founded knowledge.

(2) Technical scientists not only apply technological principles in different ways. They also, in many cases, design equipment for observation, measuring, and control. Roughly, this can be achieved by creating a connection between the relevant properties (to be called "attributes") of the donor field which is to be observed, measured, or controlled, and the properties of the recipient field, *viz.*, the new equipment.

The transference of attributes leads to ATTRIBUTIVE ANALOGY: for two objects x with attribute f, and y with attribute g, there exists a similarity (an analogy) between f and g.

The clinical thermometer is one of the most simple examples. We use the notion of *clinical* thermometer because it is meant to measure the fever of a patient. Clearly, the mercury of the thermometer itself cannot have a fever, it only takes on the same temperature as the body of the patient. So there is a similitude between the temperature of the patient and the temperature of the mercury in the thermometer; nevertheless there is also a great difference; in the one case it makes sense to speak of fever, in the other not. A similar situation occurs with regard to measuring devices for the safety of systems. It is possible to say that a "signal is unsafe," whereas in fact it is not the signal that is unsafe but the system.

This kind of analogy corresponds to medieval concepts of analogy.[3] The standard example treated by medieval philosophers is the correspondence between the meanings of the notion "healthy" as a property of food, of living beings, of complexion, of urine, etc. In the case of an *analogia attributionis* usually a causal connection is involved: the composition of the urine is caused by the state of the body, the height of the mercury in the clinical thermometer is dependent upon the temperature of the body, the pressure in a vessel changes the signal into unsafe, the kinetic energy of microscopic particles in a Glaser-chamber generates bubbles, or charged particles in vacuum tubes generate electrical pulses measured by computers.

The logic of attributive analogy is reductive logic: monitoring devices, whether for observation, measuring, or control, work satisfactorily in so far as the attribute of the donor field can be reduced to (or considered to be equivalent to) the attribute of the recipient field. This is not always the case; because of failure, safety-equipment may produce false alarm. Moreover, monitoring devices always have a limited range of application. Mercury thermometers cannot be used in all circumstances, the Glaser-chamber is not suited for the detection of, e.g., Z-particles.[4] Thus reductive reasoning involves uncertainties.

(3) From a technician one expects furthermore that he has a good grasp of the technical systems of his speciality. This is necessary for repairing, perfecting, and designing equipment. The mastery of automobile technology consists, e.g., to a large extent in understanding the system of an elementary (prototype) motor; that system is realized in different and very diverse types of motors.[5] This specific knowledge is based upon the insight into functional analogies.

We will distinguish between two types of functional analogy:

(3.1) ANAFUNC: For a layman in science and technology it will be difficult to recognize an analogy between the clutch of his motor car and a light switch. The technical scientist, however, sees in both a functional analogy, ANAFUNC, because they both perform the same kind of *function*, which moreover, in this case, can be described by a mathematical formalism (namely, network algebra). In order to avoid confusion with other types of analogies, to be discussed shortly, two things should be kept in mind. In the first place, the resemblance between the systems concerns only the technical function (and not the way the systems operate) and this resemblance is such that in principle the same formalism for describing the system can be used (in the above example, both can be described as an on-off switch). Secondly, anafunc, as introduced here, is a relation between the technical functions of two or more *objects*.

(3.2) ANASFUNC: In the anasfunc we do not consider the objects and their functions by themselves but we consider them as parts of a whole system. In contrast with the anafunc, the ANASFUNC is, therefore, a relation between *relations*. If the function of a component a of a technical system S_A is analogous to the function of a component b in a technical system S_B, we shall denote this, as is usual, in the following way:

$$a : S_A : : b : S_B$$

For a one may substitute the coupling of the mechanical system S_A and for b the switch of the electrical system S_B (see Figure 1). We find the same kind of analogy between the steering valves of the hydraulic and pneumatic systems, respectively S_C and S_D.

The anasfunc between the signal transformers consists in the fact that in all four case they function as regulators for the flow of energy and that in all four cases this energy flow can be computed in the same mathematical way, namely, as the product of a quantity representing a force and a quantity which is time-dependent.

We could prove (but will not) that both analogies are logically transitive. The importance of these proofs lies in the fact that they give a foundation to the use of mathematical formalisms, such as network algebra. In this sense our discussion of these analogies can be considered to belong to the domain of the methodological foundations of the technical sciences. Furthermore they are important for recognizing the specific properties of these analogies in comparison to other types. These analogies from technological systems theory have properties which are usually not present in the analogies used in social sciences.

(4) Besides the application of technical principles, the design of monitoring devices and the mastery of technical systems, a technical scientist must be able to construct and perform experiments on physical models and to interpret the results of these experiments in terms of the behavior of the system for which the model was constructed. A special kind of physical model is the scale model. Galileo and von Guericke were among the first in history to consider experiments on scale models.[6] Galileo discusses a machine with which ebb and flood can be imitated, and von Guericke built a sphere of sulphur, representing the earth, in order to investigate the attractive force of the earth. The use of physical models has become standard practice in aerodynamics and hydrodynamics for the design and construction of aeroplanes and ships. A very intriguing example of a physical model is the analog computer, which will be discussed more extensively later on. The use of physical models is based upon the existence of a STRUCTURAL ANALOGY between the physical system and its model; such a structural analogy exists when the physical behavior of both systems can be described by similar sets of equations.[7]

The specific properties of the structural analogy will be described in terms of the notions of structure and isomorphism. (We will use the definitions of these notions given by Bourbaki. This will be done in Section 6.)

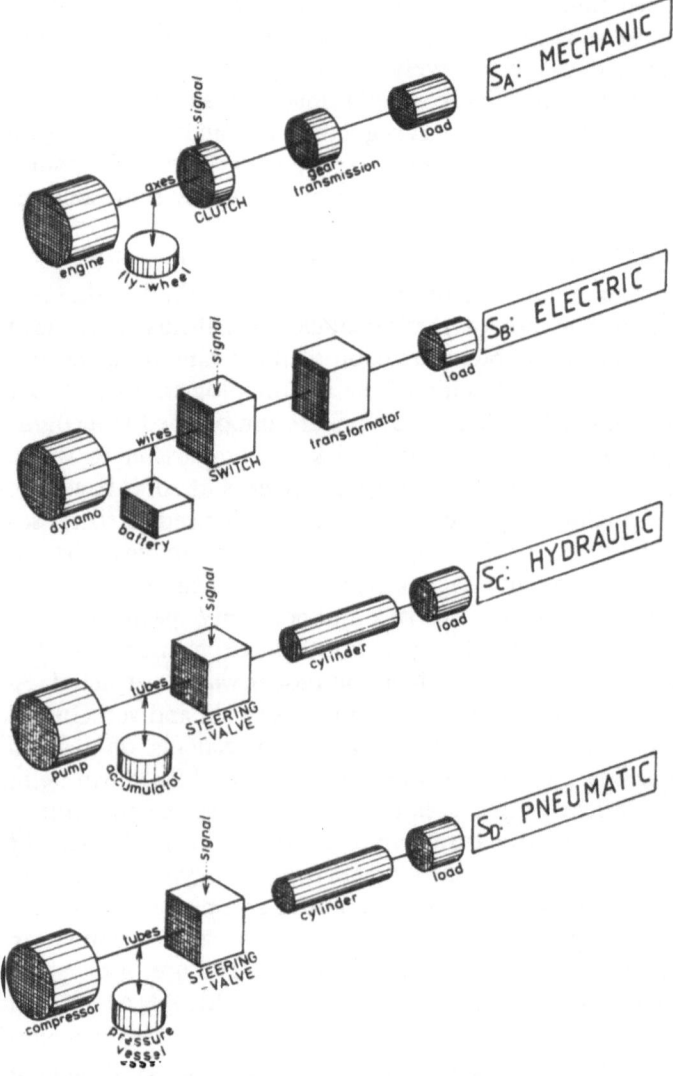

Fig. 1.   Systems with a similarity between the following elements: (1) source of energy, (2) control signal, (3) transmission, (4) load, (5) energy storage, and (6) energy transport. This drawing is taken (with some minor modifications) from J. M. Schoenmakers and H. P. Tomesen (see references below). We gratefully acknowledge permission of the NGOLB to make use of it.

2. TECHNOLOGICAL NOTIONS OF ANALOGY AND HISTORICAL TRADITIONS

In this section we shall present a sketch of the historical background of the notion of analogy in so far as it contributes to a clarification of the (differences between the) notions of analogy introduced above. We shall moreover discuss certain specific points which are connected with the fact that we are interested here in the use of analogies in the context of the technical sciences.

In Antiquity the Greeks encountered the problem (connected with Pythagoras' theorem) that it was not possible to find a common measure for certain ratios, such as $a : b = $ sqrt $2 : 1$.[8] Therefore they considered it unjustified to apply the equality sign in these cases. Instead, they assumed that there was an analogy between both cases which they expressed as follows:

$$a : b : : \text{sqrt } 2 : 1.$$

But there was still another problem as a result of which also the Aristotelian law of the lever was expressed by way of an analogy between weights and distances:

$$W_1 : W_2 : : D_2 : D_1.$$

According to the Greeks it was not permissible to equate ratios in which quantities pertaining to different domains of being appeared (such as, in the above case, weights and distances). These problems in the foundations of Greek mathematics and physics constitute the historical origin of the *analogia proportionalitatis*. Beside this type of analogy, Aristotle introduced the *analogia attributionis*; his reason for doing so was that he introduced into his philosophy of science notions like "cause," "effect," and "potency" for different sciences such as physics, biology, and astronomy, whereas the objects of these sciences belong to fundamentally different domains of being. That is why he assumed that these notions are not used in the same, but in analogous meanings in these different disciplines.

In Scholastic philosophy the proportional analogy lost its mathematical and physical meaning and became almost indistinguishable from the attributive analogy. The analogical use of "healthy" in healthy urine and healthy animal was considered to be a proportional analogy.[9] This analogy, however, has by no means the exact mathematical meaning of for instance the analogy contained in the law of the lever. We think that it is even misleading to use the same manner of writing for this medieval proportional analogy; nevertheless it is still in use in recent literature.[10] This manner of writing the analogy suggests a precision which usually cannot be given a foundation. That is why it is important to observe that in Scholastic philosophy the proportional analogy becomes a kind of attributive analogy.

It should be noted moreover that in medieval philosophy the notions of analogy were discussed within a primarily epistemological setting. The analysis of analogies was directed to the question as to how it is possible to give the same name to different attributes. This kind of problem has no longer any connection with the original Greek mathematical analogy. Apart from that, we think that it is not fruitful to analyze again ancient Greek mathematical analogy because the corresponding problems in the foundations of mathematics and physics are generally no longer considered to be relevant. On the other hand, our concept of attributive analogy does have a strong resemblance to the Scholastic attributive analogy. In our opinion, however, this kind of analogy can be used for treating the problem of theoretical concepts. (More about this will be said later on.)

The logical analysis of argumentations based upon analogies was again taken up by Hume and Kant.[11] Their main idea was that the notion of analogy could validate argumentations to a certain degree of probability. This tradition is still alive today (Nagel, 1961; Carney and Scheer, 1965; and Copi, 1982). Nagel takes over the definition of physical analogy given by Maxwell who describes it as:

that partial similarity between the laws of one science and those of another which makes each of them illustrate the other (Nagel, p. 109).

According to Carney and Scheer and Copi, and we agree with them, this problem field belongs to the domain of inductive logic.

Strongly independent of these traditions, systems theory has developed itself in the technical as well as in the theoretical sciences. Within systems theory a lot of attention has been devoted to the notion of function, but – as far as we know – no attempt has been made to make this notion more precise using the notion of analogy, as we intend to do here.

From the foregoing it becomes clear that anamorphy, attributive analogy, and functional analogy pertain to different kinds of issues because the corresponding fields with which they are related, respectively inductive logic, epistemological questions concerning theoretical notions, and systems theory, are different.

To complete this historical sketch, two more developments should be mentioned. In the first place, in the technical sciences there has already existed for a long time a strong interest in the use of physical models and there is a whole body of literature on this topic. It is known as "the theory of similitude" or *Aehnlichkeitslehre*. Important publications in this field are: Murphy (1950), Langhaar (1965), Gerhard (1971), Pawlowski (1971), and Szücs (1980). In the foregoing we have already noted that the use of physical models is based upon the existence of structural analogies.

In the second place, certain developments in modern philosophy of science are relevant. Beside the work of Nagel, which was already mentioned, the work of Mary Hesse (1966 and 1967) should be mentioned. On the one hand, she discusses intuitive ("conceptual," "material") analogies concerning which however it is very difficult to say anything precise. On the other hand, she analyzes analogies in terms of models and isomorphism; but this approach does not seem to lead to basically new results in comparison with neopositivistic philosophy of science. The neopositivists clearly were aware of the fact that a formalized theory may have different models which are isomorphic. Our approach differs from Hesse's in so far as we do not analyze structural analogies in terms of the notions of models and of isomorphism, but in terms of the notions of structure and isomorphism. Moreover, the context within which we will discuss structural analogies differs completely from Hesse's; our interest is to show how structural analogies can provide a foundation for the use of physical models in the technical sciences.

3. ANAMORPHY AND ITS INDUCTIVE LOGIC

In this section we will address the question how the use of anamorphies, as defined in section 1, can be given a foundation.

In spite of the fruitful heuristic use of anamorphies, the exploitation of them always involves uncertainties, which is characteristic of inductive logic. Arguments based upon anamorphies generally have the following form (see Copi, 1982, p. 392):

>   *a, b, c,* and *d* all have the attributes *P* and *Q*
>   *a, b,* and *c* have the attribute *R*
>   Therefore (?): *d* probably has the attribute *R*.

This shows clearly that there is always an uncertainty involved in argumentations based upon anamorphies.

Anamorphies play a role not only in heuristics, but also in the explanation of theories (whether in the presentation of the results of research, or in historical descriptions, or in the teacher/student relation). In one respect, this use of anamorphies differs from the heuristic use: the differences between both anamorphic systems are known. Usually it is required that these differences be mentioned explicitly in order to avoid a situation where the use of anamorphies obstructs the formation of new concepts.

With regard to the technical sciences and product-oriented research the following differences between the theoretician and the technician have to be kept in mind.

– Technical principles *may* be based upon theories, but this is not necessary. Technical principles and constructions may even be based upon false theoretical conceptions; for example, the Leiden jar was constructed on the basis of the assumption that electricity was a fluid.
– For a theoretician, the general validity or applicability of a theory is of paramount importance. He will generally not accept a situation in which a theory is confronted by a large number of counterexamples.[12] For a technician the general validity of his principles is not such an urgent problem. Even if a technical principle has a very limited domain of application, he will keep using it if it can be applied fruitfully. For a technician it generally does not matter that the validity of the principles of classical mechanics was restricted by the discovery of quantum mechanics; in most cases he can still apply technical principles, based upon classical mechanics, in an extremely fruitful way. Nevertheless, this difference should not be exaggerated. In solving technical problems, the technician will primarily start from known technical principles and assume their (universal) validity. This for instance was the case for Savery's, Newcomen's, and Watt's construction of the steam engine. They all used the same technical principles based upon the theory of heat of their time. Of course, Watt's condensor tripled the power of the engine, but from a physical

point of view there was no difference between the principles which he applied and those of his predecessors. In other words, there exists an anamorphism between the three constructions: the *same* principles are applied in *different* ways. From the point of view of inductive logic there is also another resemblance between the theoretician and the technician. Both do not know beforehand the limits of applicability of their principles. A nice example of technology is the history of refrigerators. In 1899 G. Linde succeeded in liquifying air with a machine which was based upon the Joule-Kelvin principle which says that the temperature of gases drops in an adiabatic expansion because of the internal work done against the mutual attraction of the molecules. In that respect his construction differed from the one of Claude which was based upon the performance of external work. Kamerlingh Onnes, however, recognized that these principles were not sufficient for liquifying gases like helium and therefore he designed a more complex liquefactor. Many other examples illustrate that again and again the limits of the applicability of technical principles has been discovered, which made it necessary to invent new ones in order to meet the requirements of the technical problems involved. In this sense there is a strong similarity between inductive logical problems connected with the application of theories and those connected with the application of technical principles. However, the "falsification" or rejection of a technical principle which has been successfully applied in different situations, is usually considered not to be such a dramatic event as the falsification of (fundamental) physical theories.

Interfering phenomena and discrepancies between obtained results and expectations play a different role in the construction of theories and in technology.[13] When predictions based upon a theory under investigation are not fully confirmed, it is possible to explain the discrepancy by making an appeal to disturbances or deviations caused by the special circumstances of the concrete situation. In such a situation it is still possible to conclude that the theory is satisfactorily confirmed by the results of measurements. In technology, on the contrary, disturbances and deviations from the intended operation of a machine are considered to be essential and usually give rise to a revision of the construction or the design.[14]

From these three points it becomes clear that a technician as well as a theoretician is confronted with inductive logical problems when applying anamorphies, albeit in different ways.

4. ATTRIBUTIVE ANALOGY (ANAT) AND ITS REDUCTIVE LOGIC

We have already stated that an attributive analogy exists when there is a connection between the attributes of a donor field (a physical or technical system) and the attributes of a recipient field (monitoring equipment for observation, measuring, or control). A precise description and analysis of the attributive analogy (ANAT) and the two functional analogies (ANAFUNC and ANASFUNC) and their differences with other types of analogies would be helpful but cannot be given here. The definition of the attributive analogy is based upon the conception of the attributive analogy as the use of the same name for two different attributes; the same name is being used because there exists a causal connection between the attributes involved. Let us add one more example from a technical context to the ones already mentioned. It is quite common to call a fuel friendly to the environment whereas in fact not the fuel but the exhaust gases are friendly to the environment; here also there is a causal connection between the composition of the fuel and of the exhaust gases.

It has to be noted that the problem of reductive logic connected with attributive analogy is not unknown to methodology. Methodologically, the logical problems concerning attributive analogy are quite similar to the problem of theoretical concepts. We shall elucidate this similarity with the help of the examples which were mentioned earlier. To that end we introduce the following symbols:

T_1 : the temperature of the body
T_2 : the height of the mercury
V_1 : the "safety" of the technical system
V_2 : the state of the signal

Now, argumentations for the determination of theoretical properties (i.e., of properties of the system to be controlled) have the following form:

Theoretical presupposition:	$T_1 \to T_2$	$V_1 \to V_2$
Observation:	T_2	V_2
Conclusion (?):	T_1	V_1

Both arguments have the same, logically problematic form. This elucidates that the ascription of theoretical properties and the introduction and use of measuring instruments is a risky undertaking from a logical point of view. In practice we are familiar with this; it is always possible that safety equipment gives off a false alarm. Moreover, on the basis of these schemata it is easy to point out the task of the technical scientist who designs monitoring devices: his designs must be such that the presuppositions in the first line of the schema can be taken to be equivalences in as many situations as possible. These equivalences, however, cannot be forced, using logical means, as Carnap once proposed. In that case the task of a technician would become very simple and could in principle be replaced by a logical postulate. From the foregoing it follows that it is useful to introduce attributive analogy into the methodology of the technical sciences.

5. FUNCTIONAL ANALOGIES: ANAFUNC AND ANASFUNC

The ANAFUNC refers to the situation in which two physical objects have the same technological function which in principle can be described with the same mathematical formalism. For reasons of simplicity we will discuss the example illustrated earlier in figure 1; as we already observed, network algebra can be used to describe the operation of the different kinds of couplings of figure 1. This use of network algebra must now be given a foundation.

Intuitively, we proceed as follows. To say of a physical system like an electrical switch that it can take the values 1 or 0 really does not make sense. On the other hand, to say that the technical function of a clutch is analogous to the one of a formal switch, is less problematic. In this manner, different physical realizations of the same function, represented by the formal switch, can be treated in the same way, i.e. within the same formalism. There are proofs which would clarify which requirements must be satisfied in order that it be possible to "jump" from the material to the formal domain, but they cannot be given here.

The ANASFUNC refers to the situation in which the function of a component a of system S_A is the same as the function of component b of system S_B. The signal transformers of figure 1 satisfy these requirements, on condition that they formally have an analogous structure. They regulate the different flows of energy which formally have the same properties. This similarity in function is represented in the figure by the fact that these regulators occupy the same "place" in the different systems. It can be shown, in a manner which strongly resembles our approach in the case of ANAFUNC, that, if these functions in the systems are conceived to be analogous, they can be considered as belonging to the same formal type.

This concludes our discussion of both types of functional analogies; we will now shift our attention from the function of a physical system to the way the system behaves physically. Similarities in the physical behavior of systems can be treated with the help of structural analogies.

6. STRUCTURAL ANALOGIES

The paradigm example of the use of a structural analogy is the analog computer. The operation of such a computer is based upon the principle that the behavior of a given system can often be simulated by constructing an electrical circuit which obeys the same differential equations as the original system. By finding a suitable correspondence between the variables and parameters of the electrical circuit, on the one hand, and the variables and parameters of the original system on the other, the behavior of the original system can be computed on the basis of the behavior of the electrical circuit.

Fig. 2a   Fig. 2b

Consider the following elementary example. The system to be investigated is a mechanical oscillator consisting of a mass m attached to a spring with force constant k (see Figure 2a). The differential equation belonging to this system is:

$$m\mathrm{d}^2x/\mathrm{d}t^2 + kx = 0$$

Now, take the electrical circuit given in Figure 2b. Its differential equation is of the same type:

$$Ld^2I/dt^2 + (1/C)\, I = 0$$

The correspondence between both systems is given by:

$$I = x$$
$$1/C = k$$
$$L = m$$

The foregoing implies that the oscillating frequency of the mechanical oscillator can be computed from the oscillating frequency of the electrical circuit

$$f = \frac{1}{2\pi}\sqrt{\frac{1}{CL}}$$

by substituting k for $1/C$ and m for L.

Of course, in this simple example, it is possible to solve the differential equation for the mechanical system and to compute the frequency of the mechanical oscillator directly; so there is no need to take the detour *via* the electrical system. For more complicated systems, however, it is often not possible to solve the differential equations; in those cases the behavior of the system can be studied by constructing and experimenting on an electrical analog.

Structural analogies are used in the technical sciences for constructing *physical models* of systems to be investigated. Here a physical model is understood to be a "device which is so related to a physical system that observations on the model may be used to predict accurately the performance of the physical system in the desired respect."[15] These physical models may be of the same physical nature as the original system, as in the case of scale models, or of a different physical nature, as in the above example. In the case of scale models, both systems are usually, but not necessarily, related by a linear transformation of the geometric variables, whereas the other relevant physical quantities of both systems are related to each other by more complex transformation rules. In finding these transformation rules dimensional analysis plays an important role. In general, dimensional analysis is applied in situations where the relevant physical variables are known, but where the differential equations describing the system are not known. In those cases experimental results on physical models can nevertheless often be evaluated with regard to the original system by using relationships between dimensionless constants.[16]

In many domains of the technical sciences, the use of physical models is indispensable for research, in particular for design and development activities. From a methodological point of view, the primary function of structural analogies lies in the fact that they enable the engineer/scientist to construct substitutes for the original system (which, for whatever reason is not available for direct investigation) on which he can perform experiments.

Let us now have a closer look at the precise nature of structural analogies and at the foundations for their use. The key to the notion of structural analogies lies in the observation that in many cases different physical systems are described by sets of mathematical equations (usually differential equations) which have loosely speaking the "same form." It is this notion, "having the same form," that needs further clarification. A precise, formal definition can be given in terms of the notions "structure" and "isomorphism."[17] Roughly, a structure is defined as a packet of relations on a set which is constructed in a particular way, and an isomorphism as a mapping from one set to another which leaves these relations invariant. Structural analogy can now be defined in the following way: two physical systems are structurally analogous when they have the same mathematical structure, i.e., whenever it is possible to construct an (iso)morphism between the mathematical structures corresponding to these systems.

This way of formalizing structural analogies has the advantage of showing clearly the justification for the use of structural analogies in technical sciences. Through the notion of (iso)morphism, the structure of a physical system (determined by the laws governing that system and boundary conditions) becomes separated from the specific nature of the physical system. The structure starts to lead its own "disembodied" life; the same structure may receive different material realizations. By concentrating on the abstract structure of physical systems, it is immediately clear that the engineer may consider one material realization of this structure as a physical model for another material realization.[18] Of course in practice the situation is usually very complex. One problem which hampers the exploitation of analogies in technology is the use of idealizations in describing the two analogical systems. These idealizations may be different for both cases. Then the difficult problem arises in what sense these idealizations affect the validity of the analogy. Besides the use of idealizations, "scale effects" may also endanger the validity of analogical reasoning.[19]

7. CONCLUDING REMARKS

The use of analogical reasoning is a common tool in science and technology, but also in other contexts (e.g., the social sciences). Very often, however, analogies are used without clearly specifying the similarities and the differences between donor and recipient field. Nevertheless this is necessary in order to avoid the danger that the notion "analogy" become synonymous with "some very vague resemblance." As we have tried to show in the foregoing, analogies in the technological sciences are not just vague resemblances, but can be given a precise meaning; there the use of analogies is an important instrument for theory formation, design, and construction. The distinction between different types of analogies, as presented here, is also relevant for the problem of the relation between science and technology. It opens new perspectives for studying the differences and similarities between the structures of thinking in science and in technology.[20]

Technical University, Eindhoven

NOTES

[1] We are grateful to A. Kamsteeg (Technical University of Eindhoven) for his assistance in the formal parts of this paper, and we would like to thank P. B. Scheurer (University of Nijmegen) for his help with the Bourbaki formalism. [**Editor's note:** The two formalisms referred to were contained in appendices which could not be printed with this article; they are available on request from the authors.]

[2] Of course, decisions on the type to which an analogy belongs are not dependent upon global descriptions of actions of technical scientists; this could lead to incompatible judgments. These decisions have to be based upon criteria derived from exact definitions of these analogies. In practice it is not excluded that several of the different types of analogy as defined here occur together in the same situation.

[3] There is a significant difference between the medieval notions of analogy and the ones introduced here; more about this will be said in the next section.

[4] The replacement of the Glaser-chamber by electronic detection with the help of computers is interesting because both methods of detection are based upon different theoretical properties of the particles.

[5] An automobile mechanic is expected to be able to repair almost any brand of automobile, even when he is confronted with a new type of motor. This expectation is based upon the presupposition that his know-how, based upon familiar motor types, can be applied analogously to new motor types. The use of failure charts in trouble-shooting is based upon these supposed analogies; see J. M. Schoenmakers and H. P. Tomesen, pp. 14–24.

[6] See Sarlemijn (1985).

[7] Note that in the case of a structural analogy the physical behavior of both systems is relevant whereas for a functional analogy only the function of the system matters.

[8] For an elaborate exposition of the role of analogy in Greek mathematics, see Szabo (1969), in particular, pp. 193–242.

[9] The Scholastics thought that the attribute "healthy" could be ascribed to the urine to the same degree as the attribute "healthy" could be ascribed to the animal.

[10] Hesse (1966, p. 62) uses expressions such as "father : children :: state : citizens."

[11] Hume (1740), toward the end of section 12. Kant (1968 [1790]), p. 594 (=B 449).

[12] Even Kuhn and Lakatos, who admit some room for counterexamples in their philosophies of science, admit this.

[13] Galileo, who intensively occupied himself with the problem of the application of "abstract" theories in concrete technical situations, had already pointed this out.

[14] For an amusing story showing that a technician somehow must react to disturbances, even if he is not able to explain his own reaction, see Casimir (1983), chap. 9.

[15] Murphy (1950), p. 57.

[16] Szücs (1980), p. 101.

[17] See Bourbaki (1966).

[18] The same could have been achieved by writing the laws of both systems in dimensionless forms. But then the problem of defining the notion of having the same form would still remain.

[19] See, e.g., Langhaar (1965).

[20] Using, with some changes, the notions of analogy presented here, Andries Sarlemijn (forthcoming) has analyzed the history of the transistor.

REFERENCES

Bocheński, J. M. 'On Analogy', *The Thomist* **11** (1948): 474–497. There is a German translation in Bocheński, *Logisch-philosophischen Studien* (Freiburg, 1959), pp. 107–129, and the article appears again in A. Menne, ed., *Logico-Philosophical Studies* (Dordrecht: Reidel, 1962), pp. 97–117 – though the latter versions differ in important details.

Bourbaki, N. *Eléments de mathématique: Théorie des ensembles*, vol. 22 (29th ed.; Paris: Hermann, 1966); see chap. 4, Structures.

Carney, J. D., and R. K. Scheer. *Fundamentals of Logic* (New York: Macmillan, 1964).

Casimir, H. B. G. *Haphazard Reality* (New York: Harper & Row, 1983).

Copi, I. M. *Introduction to Logic* (6th ed.; New York: Macmillan, 1982).

Hesse, M. B. *Models and Analogies in Science* (Notre Dame, Ind: University of Notre Dame Press, 1966).

Hesse, M. B. 'Models and Analogy in Science', in P. Edwards, ed., *The Encyclopedia of Philosophy*, vol. 5 (New York: Macmillan, 1967), pp. 354–359.

Hume, David. *Treatise on Human Nature* (London, 1740).

Kant, Immanuel. *Critique of Judgment* (1790); see *Werke*, vol. 8 (Darmstadt: Wissenschaftliche Buchgesellschaft, 1968).

Langhaar, H. L. *Dimensional Analysis and Theory of Models* (New York: Wiley, 1965).

Murphy, G. *Similitude and Engineering* (New York: Ronald Press, 1950).

Nagel, Ernest. *The Structure of Science* (New York: Harcourt, Brace & World, 1961).

Pawlowski, J. *Die Aenlichkeitstheorie in der physikalischtechnischen Forschung: Grundlagen und Anwendung* (Berlin: Springer-Verlag, 1971).

Rothbart, D. 'The Semantics of Metaphor and the Structure of Science', *Philosophy of Science* **51** (1984): 595–615.

Sarlemijn, A. 'Mechanica van 'const' tot wetenschap', in his *Van natuurfilosofie tot technische natuurkunde* (Eindhoven: Department of Physics, 1985).

Sarlemijn, A. 'Analogy Analysis and Transistor Research', *Methodology and Science* (forthcoming).

Schoenmakers, J. M., and H. P. Tomesen. 'Analogieen en systeemanalyse', in *Motorvoertuigentechniek* (NGOLB, n.d.).

Szabo, A. *Anfange der griechischen Mathematik* (Munich: Oldenburg, 1969).

Szucs, E. *Similitude and Modelling* (Fundamental Studies in Engineering, vol. 2; Amsterdam: Elsevier, 1980).

KRISTIN SHRADER-FRECHETTE

PUBLIC AND OCCUPATIONAL RISK: THE DOUBLE STANDARD

1. INTRODUCTION

According to classical economic theory, the existence of some jobs which are more hazardous than others does not constitute a social inequity. After all, movie stuntmen, coal miners, and high-rise construction workers appear to be well compensated. This means, according to the theory, that workers are free to choose among occupations so as to balance their desire for income against their desire for safety.

As might be expected, the classical economic theory allegedly supporting current patterns of occupational risk has caused a great deal of controversy. Some persons maintain that there are ethical and epistemological grounds for challenging the alleged equity of the "double standard" for public and occupational safety, while others claim that there is nothing wrong with allowing persons to use the labor market so as to trade personal safety for increased income.

Critics of classical economics and prevailing labor practice maintain that the cost of work-related injury is extraordinarily high. The National Safety Council claims that it approximates $14 billion annually, if one counts direct injuries, deaths, lost wages, administrative costs, and property losses, not to mention the human suffering.[1] Moreover, workplace and industrial casualties are statistically at least three times greater than street crime,[2] although people appear to be more concerned about the latter than about the former. Because of this fact, they argue that there should be no double standard for occupational and public exposure to various gases, chemicals, particulates, radiation, noise, and other forms of technology-induced pollution. They believe that personal safety is something which government ought to protect, and that the protection ought to be equal for everyone. On their scheme, it is not reasonable to allow certain workers to accept payment for bearing a high risk, while the public faces a lesser risk. They claim that workers ought not to have to trade their health and well-being for wages, both because such a trade attacks our basic notion of human dignity, and because it is generally questionable whether one has given informed consent to such a trade. Paying a person to put himself at risk, they say, is

257

Paul T. Durbin (ed.), Technology and Contemporary Life, 257–277.
© *1988 by D. Reidel Publishing Company.*

not essentially different from selling oneself into slavery or murder for hire.

Those who agree with the double standard for worker and public exposure to risk, however, usually maintain that the additional wages received by workers in hazardous occupations compensate them for their risks, and that government does not have the right to play "big brother" and tell workers whether or not they can trade safety for increased income. They also claim that occupational risk is overemphasized and sensationalized by the "danger establishment,"[3] and that most countries, notably the U.S., have unacceptable "rigid standards" for workplace risks. For example, a recurrent target of ridicule, for those who believe that U.S. occupational safety standards are too strict, is "the portable toilet standard for cowboys" which was set by the U.S. Occupational Safety and Health Administration (OSHA).[4]

One reason for the continuing U.S. controversy over workplace hazards, and over whether to employ a double standard for public and occupational risk exposures, is that U.S. standards for health in the workplace appear to permit greater risks than do those of many other nations. In terms of permissible levels of chemicals in the work environment, for example, U.S. regulations are less strict than those of the Federal Republic of Germany, the German Democratic Republic, Sweden, Czechoslovakia, and the U.S.S.R. Standards in Argentina, Great Britain, Norway, and Peru are approximately the same as those in the United States.[5]

The goal of this essay is to sort out the arguments for and against the claim that government ought to allow workers to accept a lower standard of health and safety in exchange for increased income. To attain this goal, the essay will proceed (1) to examine the economic theory underlying the alleged double standard for workplace and public exposure to risk; (2) to clarify and evaluate the main arguments in favor of the double standard; (3) to propose three main arguments against the double standard; and (4) to summarize the implications of all these arguments and evaluations for public policy regarding worker risk. Let us begin with the theory underlying acceptance of the double standard.

2. THE THEORY UNDERLYING ACCEPTANCE OF THE DOUBLE STANDARD

At least in the U.S., the existence of a double standard for worker and public safety is readily apparent. For example, the U.S. maximum permissible dose of whole-body radiation which can be received annually by the public is 500 millirems; the maximum permissible dose, for the same time period, for industrial workers is 5,000 millirems, or ten times as much radiation.[6]

The main reason why equity is generally not thought to demand the same standard for occupational and public exposure to various pollutants is that the two types of exposures are not thought to be analogous. According to proponents of the method of revealed preferences (for evaluating risks),[7] for example, occupational risks are usually defined as voluntary risks while public risks are defined as involuntary; since involuntarily imposed risks ought to meet more stringent safety requirements, they maintain, the double standard for occupational and public risks is reasonable.[8] Also, they claim, risks accepted "voluntarily" are more acceptable than those applied to public risks, precisely because people are compensated (through their wages) for the higher workplace risks that they bear. According to Chauncey Starr, one of the preeminent proponents of the method of revealed preferences, the risk entailed by a particular occupation is directly proportional to the cube of the wages for that occupation; as the risk increases, so do the wages.[9]

Starr's view, widely accepted among risk assessors, especially among those who follow the method of revealed preferences, is part of the classic theory of the compensating wage differential. The fundamental economic principles of this theory were formulated long ago by Adam Smith. As Smith expressed it, "the whole of the advantages and disadvantages of the different employments of labor" continually tend toward equality because the wages vary according to the hardship of occupation;[10] on this theory, men exposed to a risky workplace had advantages and disadvantages whose sum was equal to those of the men not exposed to such risks, because those in the high-risk occupations were provided with higher rates of pay than were those in low-risk jobs.[11] A great many studies have been undertaken to substantiate the existence of a pay increment for workers in riskier jobs.[12]

According to proponents of the theory of the compensating wage differential, a double standard regarding worker and public risk is acceptable because those in high-risk jobs voluntarily agree to "trade" some degree of workplace safety for higher wages. In other words, the classic solution to the problem of how to control occupational risks, and how to decide which worker risks are acceptable, is to use an "economic fix," a market mechanism, for setting standards.[13]

3. ARGUMENTS FOR THE DOUBLE STANDARD

In arguing for a market mechanism, the compensating wage differential, to resolve the problems of equity raised by the double standard for occupational and public risk, risk assessors, economists, and public policy makers generally employ at least four arguments. I will examine each of them.

3.1. *The Welfare Argument*

One approach is to use a welfare-based argument. Its proponents maintain that "insistence on uniform hazard regulations will inevitably lead to . . . detrimental" results. They claim that this is because the double standard enables those in high-risk occupations to boost "their income status above what it would otherwise have been. If all jobs were required to be as safe as the most highly paid white-collar positions, the income status of those at the bottom of the income scale would be lowered further. Wage premiums for risk do exist, but they are not sufficient to offset all of the other factors generating the low-income status of the workers who receive them." In other words, advocates of this argument maintain that the double standard for risk enhances the welfare of low-income groups because it provides them with higher wages than would a uniform standard. As Viscusi puts it, "if coke-oven workers are willing to endanger their lives in return for substantial salaries, or if India chooses to develop nuclear energy as the most promising energy source for its long-term development, government efforts to interfere with these decisions will reduce the welfare of those whose choices are regulated."[14]

Although the welfare argument is highly persuasive, in that it correctly emphasizes the importance of worker autonomy over government intervention, it is premised on a number of assumptions which are highly doubtful. Perhaps the most basic of these is that worker *preferences* are authentic indicators of desirable *values*, or at least that workers are better able to determine what is in their best interests than is government. However, in many cases, preferences are not legitimate indicators of authentic welfare, as can be seen if one examines some persons' preferences for particular marriage partners or for dangerous habits, such as smoking. It is well known to philosophers that preferences merely indicate demands, regardless of whether they are desirable demands or not, whereas welfare is concerned only with *legitimate* demands. Preferences reveal what people want. Their welfare, however, is determined by their having correct wants.[15]

Another questionable assumption of the welfare argument is that it is ethically acceptable to allow persons to trade their health and safety for money. Clearly, however, some such trade-offs would be wrong, e.g., those in which one allowed himself to be cruelly tortured in exchange for money. They might be wrong, either because they failed to acknowledge someone's rights, or because they did not respect the dignity of humans, or because they allowed the perpetrator (e.g., of the torture) to behave in reprehensible ways, or because they permitted one to use another human as a means to an end, when humans ought to be treated only as ends. In other words, it is not ethically acceptable, generally, to allow persons to trade their health and safety for money because person A's consent is not a sufficient condition for the morality of person B's actions affecting person A, even if B compensates A financially. Although they are often necessary conditions, consent and compensation are not sufficient conditions for the morality of an action, because the moral quality of an act is also determined by various rights, duties, and agreements. But if this is so, then it is not adequate to defend the theory of the compensating wage differential merely by appealing to notions of compensation, consent, or preferences.

3.2. The Market-Efficiency Argument

A second argument for accepting the theory of the compensating wage differential is that "market allocations of individuals to jobs will promote efficient matchups in many instances. If the worker bears all of the harm associated with the risk and if he is cognizant of his own particular risk, not simply the average risk for all, he will select his job optimally. . . . Workers are not in jobs at random and the market promotes the most efficient matchups."[16] For example, says Viscusi, "Blacks with the gene for sickle-cell anemia may incur a greater risk of harm from the low-oxygen conditions faced by a pilot, and female mail sorters have a greater frequency of back injuries when moving the standard seventy-pound mail sacks." If these blacks and women have accurate knowledge of the greater risks they face in particular circumstances, then they will use the market mechanism in an efficient way and will then select the job for which they are the most suited.[17]

As is probably evident, the assumptions underlying the market-efficiency arguments are quite similar to those supporting the welfare argument. Both approaches require one to assume that employees' *preferences* will operate so as to attain authentic worker *welfare*. Both contain the implicit assumption that market-based preferences are accurate indicators of legitimate values. As has already been seen, however, this assumption is not generally true. If it were, there would never be grounds for government intervention in markets, e.g., to set minimum standards for workplace conditions. Likewise, were this assumption true, then one would have to condone the sweatshop conditions of a century ago. One would have to agree that twelve-hour workdays of a bygone era were efficient, because they allowed workers to choose an "efficient matchup." On the contrary, the efficiency and the optimality of worker choices, whether among anemia-prone blacks or backache-prone women, is in part a function of the choices *available* to workers. If an economy is not diversified, and if employees have no real occupational alternatives in the face of the need to feed their families, then it can hardly be said that the "market . . . will promote efficient match-ups."

The market-efficiency argument is also highly questionable in that the ethical conditions necessary for desirable market transactions are frequently not met. Recall that Viscusi maintained that (italics mine): "*If the worker bears all of the hazard associated with the risk and if he is cognizant of his own particular risk*, not simply the average risk for all, he will select his job optimally" with respect to his own risk potential and personal advantages and disadvantages. This means that, on its proponents' own terms, the validity of the market-efficiency argument is premissed on workers' having adequate knowledge of their particular risk situations. But are people generally aware of their own risk potential? Most risk assessors would probably say that they are not. Starr, Whipple, and other proponents of the method of revealed preferences, as well as Fischhoff, Slovic, Lichtenstein, and other advocates of the method of expressed preferences (see note 7) have pointed out, repeatedly, that intuitive or subjective assessments of risks made by educated laymen are quite divergent from analytical, allegedly objective assessments of risks made by risk experts. Laymen typically overestimate low-probability risks and underestimate higher-probability ones. For example, they overestimate catastrophic nuclear accident risks, but underestimate risks associated with automobile accidents.[18] Economists also realize that the public's risk perceptions are rarely accurate. To bridge the gap between the theoretical model of rational choice and actual, imperfect, real-life choice, economists almost always write the costs of searching for risk information into their equations dealing with choice under uncertainty. If these economists and risk assessors are correct, then the conditions necessary for ethical use of the argument from market efficiency are frequently not met. But if these conditions are not satisfied, then the argument does not provide convincing grounds for supporting the theory of the compensating wage differential.

3.3. *The Autonomy Argument*

A third reason for risk assessors' supporting the theory of the compensating wage differential is their allegation that it provides for more worker freedom and autonomy than would a theory not based on a monetary differential, but based instead on uniform standards. As one proponent of the autonomy argument puts it: "Uniform standards do not enlarge workers' choices; they deprive workers of the opportunity to select the job most appropriate to their own risk preferences," and they enable rich persons to impose their risk preferences on lower income classes.[19] On this theory, acceptance of uniform risk standards and rejection of the compensating wage differential are not desirable because they represent "interference with individual choices."[20]

Like the previous two arguments, this one is also based on the doubtful presupposition that freedom and autonomy are served by identifying occupational *preferences* with authentic worker *welfare*. As has already been noted, such an identification does not work in all cases. The presupposition also fails to take account of the fact that, just because one holds a particular job, this does not mean that his occupation is an expression of his preferences. Many people engage in a certain work, not because they freely and autonomously choose to do so, but because they have no other alternatives. Moreover, in the absence of minimum standards for occupational safety, and in the absence of alternative opportunities for employment, one could hardly claim that his occupation was a result of autonomous choice. In fact, minimum risk standards or stricter safety requirements might actually enhance occupational autonomy, in the sense that workers might not be forced by circumstances to accept jobs whose risks were higher than they wished to bear. In failing to take account of the numerous factors which limit free choice, Viscusi and other proponents of the autonomy argument appear to assume, erroneously, that government safety regulations always limit workers' freedom, and that these alleged limitations are more significant than those imposed by lenient standards governing occupational safety.

3.4. *The Exploitation Avoidance Argument*

Many proponents of the theory of the compensating wage differential realize, however, that occupational safety and worker welfare are not always guaranteed simply by letting market forces operate. They know that often employees can be exploited by employers who are not forced to provide a safe working environment. To counteract this tendency, some proponents of the theory of the compensating wage differential maintain that a necessary condition for ethical implementation of this theory is that workers have adequate information about the risks they incur. According to Viscusi, "the most salient" form of market failure is inadequate worker information. "If workers and firms are not fully cognizant of the job risks resulting from their decisions, the desirable properties usually imputed to market outcomes may not prevail."[21] To avoid worker exploitation and market failure of the theory of compensating wage differentials, proponents of the theory often advocate employee education. Their view is that, once worker education is adequate, then market forces will drive compensating wage differentials so that optimal match-ups between employees and occupations will occur.

Admittedly this exploitation avoidance argument is an improvement over arguments which ignore the role of occupational risk education but which support the theory of the compensating wage differential. Its flaw is in its major presupposition that education and compensation, alone, provide sufficient grounds for worker consent and autonomy. It takes too simplistic a stance as to the requirement for legitimate consent and free choice. As the torture example cited earlier reveals, other factors, beside one's knowledge of a situation and his being compensated for losses, determine the moral quality of his and others' choices about that situation. As was already noted, even a perfectly informed worker, who consented to the level of compensation for his high-risk job, nonetheless might have been forced to take the work, particularly if there were no alternative employment opportunities available or if he needed the money. This suggests that, in addition to workers' having full knowledge of their risk situation and being compensated for it, occupational choices must also be made in a context of ethically desirable background conditions. Such background conditions might include the operation of a free market and the existence of alternative employment opportunities. Without these background conditions, it is not clear that ethically desirable employee-employment match-ups will occur.

Take, for example, the ethical desirability of choices made by miners who choose to work in Appalachian coal fields. (Appalachia includes much of the states of Kentucky, West Virginia, Virginia, Tennessee, North Carolina, and South Carolina.) It is well known that mining is one of the highest-risk occupations, that poorer workers are typically employed in the most risky jobs,[22] and that residents of Appalachia generally have no alternative to working in the mines, unless they want to move out of the region. This is because the Appalachian economy is not diversified, because there is no job training in a variety of jobs, and because absentee corporations who control eighty percent of all Appalachian land and mineral rights also control the only jobs; the situation is one of monopsony, where owners of most of the land also control most of the jobs.[23]

Even if Appalachian coal miners were generously compensated, and even if they all had perfect information as to the dangers of their jobs, the background conditions in the Appalachian economy would prevent their making a wholly voluntary choice to work in the mines. But if they were not able to make wholly voluntary choices as to the form of their employment, then it is not clear that proponents of the theory of the compensating wage differential can argue either that, since workers were aware that their jobs were extremely risky, those risks were freely chosen, or that the prevailing double standard with respect to occupational and public risk is acceptable to workers. In fact, if background conditions necessary for procedurally just choices (about forms of employment) are not met, it is not clear that implementation of the theory of the compensating wage differential is just. As John Rawls put it, "Only against the background of a just basic structure . . . and a just arrangement of economic and social institutions, can one say that the requisite just procedure [for occupational and other choices] exists."[24] Sound as this insight about background conditions is, many risk assessors often neglect it in their considerations. In an otherwise excellent book on risk, even the brilliant philosopher Nicholas Rescher appears to neglect the role of background conditions in determining ethically acceptable risk choices. He speaks, for example, of suicide as being a "wholly voluntary" mode of death and of incurable disease as being a "wholly involuntary" mode of death.[25] Such language, however, ignores the importance of background conditions in determining what is more or less voluntary. Death by suicide might not be "wholly voluntary" (as he says) if it is a consequence of depression-induced medication whose side effects were unknown by the patient and by the doctor prescribing it. Likewise, death by incurable disease might not be "wholly involuntary" (as he says) if it is brought on more quickly by a person's unwillingness to take proper medical treatments, follow prescribed diets, etc. In other words, the line between what is voluntary and involuntary is quite uncertain in numerous cases. To the degree that philosophers and risk assessors ignore the numerous ways in which background conditions can affect the voluntariness of an action, to that same extent are they also likely to misjudge the voluntariness with which persons (e.g., Appalachians) choose to accept a particular level of risk. And to the degree that they misjudge voluntariness, they are also likely to propose inadequate theories about the ethics of risk acceptability.

In addition to the Appalachian example, there is further evidence for the thesis that, even with full information about risk, workers often are unlikely to make wholly voluntary decisions to accept high-risk employment situations. This is that people who can afford to do so generally avoid working in hazardous occupations. It is well known that, as a person's income increases, his willingness to accept risky situations decreases.[26] If this wealth-risk relationship holds, then workers' acceptance of high occupational risks is explicable by the constraints imposed by their low income and limited job skills, regardless of whether they understand the dangers to which they are exposed or not.

Even if proponents of the exploitation avoidance argument are correct in believing that proper education of workers can theoretically block exploitation of employees in high-risk occupations, it is still not clear that, practically speaking, such education can be accomplished to the degree necessary in all situations. In other words, even if education were a sufficient condition for insuring that high-risk workers voluntarily accepted the terms of their employment, it is not clear that this condition could be met in most situations. Hence, it is not clear that one would be justified in implementing a system of compensating wage differentials.

Why might the condition on education not be met? One reason is that either deliberately or out of negligence, companies and regulators have often kept their research findings about hazards secret from employees exposed to them. In the case of vinyl chloride, for example, long before workers were discovered to be at risk from liver cancer, there was strong enough evidence to support a presumption of a serious occupational hazard. Similarly, decades after countries such as Japan have banned certain carcinogenic dye ingredients from the workplace, American workers "are still literally sloshing in them."[27] When company doctors have been aware of employment-induced illness, e.g., from asbestos in the Johns-Manville factory in Pittsburgh, often they have covered up this fact for decades.[28]

Even some proponents of the compensating wage differential point out that "available evidence suggests that few firms make a comprehensive effort to inform workers of the risks they face. For example, no firms tell their employees the average annual death risk they face. Much information that firms do provide is not intended to enable workers to assess the risk more accurately. . . . Rather, it is directed at lowering workers' assessments of the risk. The most widespread claim by firms is that National Safety Council statistics indicate that the worker is safer at work than at home – a statement that . . . is intentionally misleading [because some jobs are riskier than the average home, while others are not]."[29]

In situations where there is no deceit on the part of employers regarding the relevant risks faced by their employees, and in which workers are provided with full information, even this is not enough to insure that the practical conditions necessary for wholly rational occupational choices have been met. Proponents of the method of expressed preferences have pointed out that, even in the presence of complete disclosure on the part of the company, employees exposed to high-risk situations typically take on the "it won't happen to me syndrome."[30] The pervasiveness of this syndrome indicates that, even when the theoretical conditions for full employee education are met, they might not be satisfied in a particular concrete case, owing to misperception on the part of the worker. This in turn means that, because their knowledge is not operative, many employees likely are not making wholly voluntary decisions to work in high-risk situations.[31]

4. ADDITIONAL ARGUMENTS AGAINST THE DOUBLE STANDARD

In addition to these considerations that full education and compensation do not constitute sufficient conditions for affirming that employees in high-risk occupations accept their jobs in a fully voluntary sense, there are several other reasons why one might argue that the currently accepted theory of the compensating wage differential and the double standard is not necessarily ethically defensible. One of these arguments is epistemological, and the other two are ethical. Let us consider the epistemological argument first.

4.1. *The Argument That There Is No Compensating Wage Differential*

The Achilles heel of the claim, that the double standard for public and occupational risk exposure is justified, is the assertion that there is a compensating wage differential. Despite ·the data that have been amassed to exhibit the existence of the differential (see notes 11 and 12), there are strong grounds for doubting that it exists in any general sense. The most important of these grounds is the charge that wage and risk data have been aggregated in a particular way; when they are disaggregated, the alleged differential disappears.

In a classic study on the compensating wage differential, Graham and Shakow showed the effects of disaggregating the wage and risk data. When all workers were lumped together according to salary, from lowest paid to highest paid, then the average annual probability of fatality could be shown to increase as the average worker salary increased, much as the theory of the compensating wage differential predicts; each increment in risk of death could be shown to be associated with an increment in pay. When Shakow and Cyr followed labor market segmentation theory, however, and disaggregated the workers into two separate groups, then they were able to show that, despite higher risk, compensation did not exist at all for many subgroups in the labor force.[32]

Graham and Shakow divided workers into either primary or secondary segments on the basis of a number of criteria. The primary group included skilled workers, males, workers in monopoly-sector industries, unionized workers, workers in groups with seniority benefits, and workers with high-stability jobs and high wages. The secondary group, on the other hand, included unskilled workers, females and members of minority groups, workers in small firms, non-unionized workers, workers without seniority benefits, and workers with unstable jobs and low wages. As one might suspect, the male, unionized, and skilled primary group included many white-collar workers, whereas the female/ minority, non-unionized, and unskilled secondary group included primarily blue-collar workers. Graham and Shakow discovered that, although there is some evidence for a compensating wage differential within the primary group, there is no evidence for such a differential in the secondary group, even though members of the secondary group bear risks equal to, or greater than, those in the primary group. Instead, the differential only appears to exist because, when the primary and secondary groups are lumped together, the differential of the primary group averages out among both groups and gives the appearance of a differential existing for everyone.[33]

If Graham and Shakow are correct, then the alleged compensating wage differential does not exist for the large sector of workers that are female, members of minority groups, unskilled, non-unionized, and in small firms. This means that the very persons who have the most to gain from the existence of a compensating wage differential are precisely those for whom it probably does not exist. And if it does not, then any argument that the market, through such a differential, compensates persons for their risks is not generally true. This means that, before anyone appeals to the existence of a compensating wage differential in a particular case, he will need to examine labor market segmentation theory to determine whether the differential even exists for the group in question.

4.2. Acceptance of Some Risks Involuntarily Imposes Them on Others

Consider the case in which a worker allegedly accepts a high workplace exposure to some carcinogen in exchange for a very high wage differential. The employee might be fully cognizant of the health hazards involved, and he might agree that the compensation afforded him is adequate. Nevertheless, he might not have the right to take the risk, perhaps because his acceptance of it inevitably puts other persons, who have not agreed to accept the risk, in jeopardy. Since most carcinogens are also mutagens, the worker who exposes himself to carcinogenic materials is also likely exposing his potential children and their descendants to mutagenic hazards. Of course, one might argue that unborn members of future generations have no rights to be protected from mutagenic risks taken by their ancestors, since they are not existent, and only existent beings have rights.

While the whole issue of rights of future generations is too extensive a topic to be discussed here,[34] one fact about the carcinogenic/mutagenic risk situation does seem clear. Workers might have a right to take a risk which endangers themselves. It is less obvious that they have a right to take a risk which might damage something, the gene pool, which is beyond themselves. Hence it could well be questionable to assert that any person ever intending to reproduce has the right to accept work-place risks which are mutagenic when those risks are higher than those to which the public is normally exposed. As Rescher puts it so well: we can only take risks for ourselves, not for others; "morality enjoins conservatism."[35] Or, more strongly, the moral aspect of risk taking arises when the choices of individuals bear upon the interests of others.[36]

One does not have to move to future generations, of course, to discover innocent victims of some worker's alleged right to expose himself to industrial toxins in exchange for a higher wage. For example, some occupations, e.g., that of air traffic controller, produce high psychological risks, e.g., that of stress. It might be questionable whether someone has the right to accept such a high-stress risk when the effects of the stress are not borne merely by the employee but also by his family. Likewise it might be questionable whether a particular worker, for example, in an asbestos factory, has the right to accept a higher workplace risk if such a risk might also affect his family. It is common-place for the families of particular workers to contract cancer because they have been exposed to asbestos fibers carried home on clothing. Some wives have died of asbestos-induced cancer merely because they washed their husbands' clothing. Close contact with their fathers has also caused the children of asbestos workers to contract cancer, and recent U.S. examinations have revealed dangerous levels of lead in the blood of lead workers' children, chiefly as a consequence of inhaling lead dusts brought home on clothes.[37]

Admittedly some of the hazards faced by families of those in high-risk occupations could be eliminated or reduced by simple practices such as workers' bathing and discarding their work clothes before coming home. Nevertheless, to the extent that any employee's acceptance of a risk thereby places a higher health risk on someone other than himself, then to that same degree is his right to take such a risk highly questionable.

4.3. *Acceptance of Some Risks Is Based on Inconsistent Attitudes about Risk Perception*

A second reason why one might argue that a compensating wage differential does not justify societal acceptance of workplace risks which are higher than public risks is that proponents of the differential often defend their stance by making inconsistent appeals to the status of risk *perceptions* of workers. When Starr and other proponents of the theory of the compensating wage differential wish to justify workers' acceptance of higher risks in return for higher wages, they take an interesting stance. They maintain that, once employees are adequately educated regarding the risks they face, their risk *preferences* ought to be followed, and that the regulators have no right to tell workers that they cannot follow their preferences for higher risks.[38] However, when these same proponents of the theory of the compensating wage differential wish to justify government imposition of particular standards for public risk, in the face of citizens' demands for stricter regulations, they take a different stance. They maintain that the risk preferences, even of highly educated laymen, are subjective, intuitive, and generally erroneous, and therefore that regulators ought not merely follow the public's demands for lower risks. Instead, they claim, regulators also ought to adhere to the risk assessments calculated by experts, since these reflect "rational" preferences, and they ought to implement them in standards for public exposure.[39]

For example, speaking of the public's "irrational" aversion to low-probability, high-consequences nuclear accidents, Starr and Whipple maintain that laymen's perceptions regarding this technology are incorrect and that their demands for greater nuclear safety are not reasonable, since they fly in the face of experts' beliefs about acceptable levels of nuclear risk.[40] Psychometric surveys of attitudes about risk reveal that there is no significant difference, in level of relevant technical knowledge, between those who favor greater nuclear safety for the public and those who favor less stringent nuclear risk standards.[41] Nevertheless, Starr and others claim that the preferences of the public for lower nuclear risks ought not to dictate risk standards.[42]

Proponents of the theory of the compensating wage differential, who *advocate* adherence to worker perceptions of risk, in order to justify *less stringent occupational standards*, are thus in an apparently contradictory position when they *condemn* adherence to risk perceptions of relevantly educated laymen, in order to justify laypersons' proposals for *more stringent public standards*. They cannot have it both ways. Either government standards ought to be based on the risk perceptions of adequately informed persons who are likely to be affected by the risk, or they ought not to be based on these risk perceptions. If risk assessors claim that relevantly educated persons err in their risk perceptions and ought to be "corrected" by experts, then both workers and the public ought to be so corrected, and not just the public.

Of course, the main objection to this appeal for consistency, in valuing the risk perceptions of those who are adequately informed about a particular hazard, is that the cases of worker perceptions and public perceptions are not analogous. One might argue, following this line of objection, that workers voluntarily accept given modes of employment and that they have differential compensation for the risks they face, whereas the public has neither. Because of the alleged consent and compensation involved in the worker case, so the objection goes, workers' preferences ought to be followed, whereas risk preferences of the public (which generally involve cases of less consent and no compensation) need not be followed.

As this objection correctly notes, the cases of workers' perceptions and the public's perceptions are disanalogous with respect to consent and compensation. It does not follow, however, that these disanalogies are morally relevant in justifying inconsistent treatment of risk perceptions. Why not? If the fact that one is compensated for the risk he chooses is reason to follow his risk preferences, as proponents of the compensating wage differential argue, then the fact that one is *not* compensated for the risk imposed on him is even greater reason to follow his preferences for lower risks. Virtually all risk assessors, especially those who adhere to the method of revealed preferences, including Starr and other advocates of the compensating wage differential, maintain that risks taken voluntarily are more acceptable than risks of the same level which are involuntarily imposed.[43] But if this is so, then there is greater reason to follow public preferences for lowering risks to which citizens are involuntarily exposed than there is for following worker preferences regarding risks for which consent is not necessarily wholly voluntary (see Section 3.4 in this essay). In other words, the very disanalogies between worker risk and public risk, with respect to compensation and consent, indicate that, if anything, there is very likely more reason to follow public preferences for lower risks than to follow worker preferences for higher risks (since the public is neither compensated for societal risks nor given a choice as to whether to accept them, and since workers' acceptance of jobs is often not voluntary, owing to questionable background conditions). Yet, this is exactly the *opposite* of the view taken by most risk assessors, who argue against following societal preferences for lower risks and in favor of worker preferences for higher risks. This means that proponents of the compensating wage differential are on shaky ground when they both reject and accept risk preferences, depending on whether those preferences emerge from the public or from workers. If Starr, Whipple, and other risk assessors are correct in rejecting public preferences about societal risks, then it is highly questionable to invoke worker preferences in order to help justify the theory of the compensating wage differential.

5. CONCLUSIONS AND ALTERNATIVES

If these responses to the arguments in favor of the theory of the compensating wage differential are correct, then it is likely that appeal to this theory is not adequate grounds for defending a double standard with respect to occupational and public risks. Compensation and allegedly voluntary choice of occupation do not guarantee that a particular level of worker risk is ethically acceptable, any more than compensation and the subject's consent guarantee that other actions affecting a subject are ethically acceptable. As was already pointed out, if a particular action is wrong, e.g., engaging in nontherapeutic experimentation on human beings, then the fact that the human beings may have consented to the experimentation and that they may be compensated for it, does not change the ethical quality of the act (of experimentation) from "undesirable" to "desirable." Admittedly, however, consent and compensation might render the act less undesirable than it might have been, were the subject not to have given consent or not to have received compensation. In other words, the whole question of whether a double standard for occupational and public risk is morally acceptable cannot be reduced simply to the issue of compensation and consent, as proponents of the theory of the compensating wage differential appear to do.

If compensation and consent are not the only relevant considerations in deciding whether the double standard for occupational and public risk is acceptable, then the theory of the compensating wage differential, alone, does not provide grounds for accepting such a double standard. This means that, in the absence of some ethical justification for the double standard, the best policy might be to follow a principle of *prima facie* equality.[44] In other words, in the absence of good reasons for discriminating with respect to equal protection, the best policy might be to protect persons equally, whenever possible, regardless of whether they are workers or members of the public.

If it turns out that there are compelling reasons, other than the theory of the compensating wage differential, for continuing to maintain a double standard with respect to occupational and public risk, then those reasons will need to be defended. One place to begin investigating whether there are good reasons for maintaining a double standard might be to think of worker risk as analogous to patient risk. Although there is an ethical and a legal requirement for informed consent on the part of a patient being treated by a medical doctor, one of the limitations of the current policy of following the theory of the compensating wage differential is that there is no legal requirement for informed consent in the workplace. One might well argue, however, that just as persons now claim that a doctor's witholding information from a patient is a violation of the medical doctor's fiduciary role and a way of undermining the patient's autonomy, so also it might be said that an employer's witholding of risk information from employees is a violation of the employer's fiduciary role and a way of undermining the employees' autonomy. Were there a recognized ethical and legal requirement for informed consent in the workplace, then the case for the ethical desirability of the compensating wage differential would be much stronger. Indeed, with an adequate means of insuring informed consent, and with a situation of just background conditions, there might be ethical grounds for justifying a double standard with respect to occupational and public risk.

Regardless of possible future justifications of the current double standard for risk, one thing is certain. The theory of the compensating wage differential, as now implemented, does not adequately safeguard worker autonomy and well-being, for all the reasons spelled out earlier. This means, at a minimum, that risk assessors ought either to accept uniform risk standards, on grounds of *prima facie* equality, or to provide further ethical justification for their current adherence to double standards. If risk assessors do neither, then they will be open to the charge that they have allowed market expediencies, rather than ethical convictions, to dictate standards for worker health and safety.

University of South Florida

NOTES

[1] J. B. Revelle and Robin Bates, 'OSHA – Still Not Enough', *Professional Safety* (April 1977): 42.

[2] Carl Gersuny, *Work Hazards and Industrial Conflict* (Hanover, N.H.: University Press of New England), p. xi; hereafter cited as: *WHIC*.

[3] M. Douglas and A. Wildavsky, *Risk and Culture* (Berkeley: University of California Press, 1982), p. 9; hereafter cited as: *RAC*.

[4] W. K. Viscusi, *Risk by Choice* (Cambridge, Mass.: Harvard University Press, 1983), pp. 114–115, 136; hereafter cited as: *RBC*. For an excellent treatment of the history of occupational risk and disease, see D. M. Berman, *Death on the Job* (London: Monthly Review Press, 1978); hereafter cited as: *DOJ*. See also the numerous case studies in L. B. Lave, ed., *Quantitative Risk Assessment in Regulation* (Washington, D.C.; Brookings Institution, 1982), chapters 3–8; hereafter cited as: *QRA*.

[5] R. W. Kates, *Risk Assessment of Environmental Hazards* (New York: Wiley, 1975), pp. 46–47; hereafter cited as: *RA*.

[6] *Code of Federal Regulations*, 10, Part 20 (Washington, D.C.: U.S. Government Printing Office, 1986), pp. 253ff.

[7] See sections 3.32–3.33 chapter 2, K. S. Shrader-Frechette, *Risk Analysis and Scientific Method* (Dordrecht: Reidel, 1984); hereafter cited as: *RASM*.

[8] For this point of view, see C. Starr, 'Social Benefit Versus Technological Risk', *Science* 165/3899 (19 September 1969): 1232–1233; hereafter cited as: Social Benefit. See also N. Rescher, *Risk: A Philosophical Introduction* (Washington, D.C.: University Press of America, 1983), p. 172; Rescher argues that involuntary risks are less acceptable and hence ought to be subject to more stringent standards; hereafter cited as: *Risk*.

[9] Starr, 'General Philosophy of Risk–Benefit Analysis', in H. Ashley, R. Rudman, and C. Whipple, eds., *Energy and the Environment* (New York: Pergamon, 1976), p. 16; hereafter cited as: Philosophy and *EAE*. See note 10 above.

[10] Viscusi, *RBC*, p. 38.

[11] Viscusi, *RBC*, pp. 37 ff., 156–168. See also Douglas MacLean, 'Risk and Consent: A Survey of Issues for Centralized Decision Making', working paper, Center for Philosophy and Public Policy, University of Maryland, College Park, Maryland, 1981, pp. 6–9; MacLean refers to the theory of compensating wage differentials as part of what he calls the "model of implied consent."

[12] See, for example, C. Brown, 'Equalizing Differences in the Labor Market', *Quarterly Journal of Economics* 94 (February 1980): 113–134; A. E. Dillengham, *The Injury–Risk Structure of Occupations and Wages*, Ph.D. Dissertation, Cornell University, Ithaca, New York, 1979; R. A. McLean et al., 'Compensating Wage Differentials for Hazardous Work: An Empirical Analysis', *Quarterly Review of Economics and Business* 18, no. 3

(1978): 97–107; C. Olson, *Trade Unions, Wages, Occupational Injuries, and Public Policy*, Ph.D. Dissertation, University of Wisconsin, Madison, 1979; R. S. Smith, 'Compensating Wage Differentials and Hazardous Work', Technical Analysis Paper No. 5, Office of Evaluation, Office of the Assistant Secretary for Policy Evaluation and Research, U.S. Department of Labor, Washington, D.C., 1973; R. Thaler and S. Rosen, 'The Value of Saving a Life', in N. E. Terleckyi, ed., *Household Production and Consumption* (New York: National Bureau of Economic Research, 1976), pp. 265–298; and W. K. Viscusi, *Employment Hazards* (Cambridge, Mass.: Harvard University Press, Cambridge, 1979).

[13] Viscusi, *RBC*, pp. 46, 52.

[14] Viscusi, *RBC*, p. 52. See also Wildavsky and Douglas, *RAC*, pp. 69 ff.

[15] See Shrader-Frechette, *RASM*, Chapter 2, section 3.33.

[16] Viscusi, *RBC*, pp. 132, 135.

[17] Viscusi, *RBC*, 132–133.

[18] See, for example, C. Starr and C. Whipple, 'Risks of Risk Decision', *Science* **208**/4448 (6 June 1980): 1115–1117; hereafter cited as: *Risk*. Finally, see also B. Fischhoff, P. Slovic, S. Lichtenstein, S. Read, and B. Combs, 'How Safe is Safe Enough?' *Policy Sciences* **9**, no. 2 (1978): 140–142, 148–150; hereafter cited as: Fischhoff, Safe.

[19] Viscusi, *RBC*, p. 80.

[20] Viscusi, *RBC*, p. 83.

[21] Viscusi, *RBC*, p. 76; see also pp. 77–87.

[22] See, for example, M. W. Jones-Lee, *The Value of Life: An Economic Analysis* (Chicago: University of Chicago Press, 1976), p. 39; Eckholm, Jobs, pp. 33–34. See Starr, Philosophy, pp. 15 ff., and Viscusi *RBC*, p. 46.

[23] John Egerton, 'Appalachia's Absentee Landlords', *The Progressive* **45**, no. 6 (June 1981): 43–45, and J. Gaventa, W. Horton, and the Appalachian Land Ownership Task Force, *Land Ownership Patterns and Their Impacts on Appalachian Communities*, vol. 1 (Washington, D.C.: Appalachian Regional Commission, 1981), pp. 25–59, 210–211.

[24] J. Rawls, *A Theory of Justice* (Cambridge, Mass.: Harvard University Press, 1971), p. 87.

[25] Rescher, *Risk*, p. 173.

[26] See, for example, B. A. Emmett, *et al.*, 'The Distribution of Environmental Quality', in D. Burkhardt and W. Ittelson, eds., *Environmental Assessment* (New York: Plenum, 1978), pp. 367–371, 374; and P. S. Albin, 'Economic Values and the Value of Human Life', in S. Hook, ed., *Human Values and Economic Policy* (New York: New York University Press, 1967), p. 97. See also M. Jones-Lee, *The Value of Life* (Chicago: University of Chicago Press, 1976), pp. 20–55.

[27] Eckholm, Jobs, p. 33.

[28] Berman, *DOJ*, pp. 1–4.

[29] Viscusi, *RBC*, p. 71.

[30] Starr, Philosophy, p. 5.

[31] See Viscusi, *RBC*, pp. 60–75.

[32] J. Graham and D. Shakow, 'Risk and Reward', *Environment* **23**, no. 8 (October 1981): 14–20, 44–45; J. Graham, *et al.*, 'Risk Compensation', *Environment* **25** (January/February 1983): 14–27.

[33] See note 32.

[34] For discussion of this topic and relevant bibliographical materials, see K. S. Shrader-

Frechette, *Environmental Ethics* (Pacific Grove, Calif.: Boxwood Press, 1981), Chapter 3.
[35] Rescher, *Risk*, p. 161.
[36] Rescher, *Risk*, p. 162.
[37] Eckholm, Jobs, p. 30.
[38] Viscusi, RBC, pp. 77, 80, 83. See Starr, Social Benefits, pp. 1233–1234, and Starr, Philosophy, pp. 15–21.
[39] Starr and Whipple, Risks, pp. 1115–1119.
[40] Starr and Whipple, Risks, pp. 1115–1117, esp. p. 1117.
[41] Fischhoff, Safe, p. 150.
[42] See Shrader-Frechette, *RASM*, Chapter 6, section 4.
[43] Starr, Social Benefit, pp. 1233–1234; Starr, Philosophy, pp. 26–30.
[44] See Shrader-Frechette, *RASM*, Chapter 3, for a discussion of *prima facie* equality.

WALTHER CH. ZIMMERLI

VARIETY IN TECHNOLOGY, UNITY IN RESPONSIBILITY?

I

The contemporary ethical debate within the framework of the philosophy of technology, as long as it is not a matter of mere statements of will but of rational arguments, is characterized by two convictions:

1. Almost all who write on the subject agree that, within the framework of discussion of new forms of technology, ethics should play a decisive role.
2. Almost all agree that the new ethical climate must assume, with Max Weber (1964, pp. 174f), that it is not a matter of deontological ethics, but of a teleological ethics of responsibility (Jonas, 1979).

I am of the opinion that Weber's distinction has been superseded by contemporary developments. The alternative is no longer deontological ethics or an ethics of responsibility. It has long since become a deontological question whether we should practice an ethics of responsibility. On the contrary, the decisive problem appears to me to consist in a contrast: as far as technological development is concerned, *change* functions as a premise; whereas what we call "responsibility" is usually thought of in a *static* fashion. In what follows, it will therefore be necessary to inquire into the relationship between technological development and the development of the concept of responsibility.

In order to achieve this goal, I take three steps. After a few introductory remarks on the relationship between the most significant concepts – "man," "technology," "nature," and "culture" – I shall (1) undertake an ideal-typical reconstruction of the phases of technological development. Against this background (2) the question will then be raised as to whether the concept of responsibility has developed correspondingly – and what changes must be registered in it if both technological development and the concept of responsibility are to be related to one another. On the basis of these analyses (3) I will then be able to discuss the specific requirements of a concept of responsibility in the face of post-modern technology.

The difference between the past and the present is a natural

279

Paul T. Durbin (ed.), Technology and Contemporary Life, 279–293.
© *1988 by D. Reidel Publishing Company.*

prerequisite for our subjective sense and measuring of time. On the other hand, this holds good only on the condition that, in spite of everything, something remains identical from past to present. Asking about changed technology assumes an identical core of changing technology. The thesis with which I should like to begin is a variant of this assumption:

(I) *Although, with our new technology, we are clearly already moving into a post-modern age, our thinking about nature and technology still gets its orientation from a traditional-substantialist model. We imagine technology as a substance which remains while its qualities change. Such a conception only leads to a pointless "quarrel over technology"* (Dessauer, 1956) *which in fact becomes a quarrel over words and definitions. Instead, we should take into account that technology, man, nature, and culture are only nodal points of relations in a tight network. Although these relations can be defined functionally, it is the relationships that remain structurally constant, not the nodal points. The individual elements change historically according to a definite pattern, whereas the function of the relation remains unaffected. I define this relationship as follows. By the utilization of tools man changes nature so as to be able to acquire both nature's material and immaterial values; "culture" is the different social and historical ways in which this change occurs.*

The distinguishable historical stages in the recent history of mankind are characterized by specific profiles corresponding to these relationships. (This will be clarified in what follows.) Ideally, I distinguish four different stages (Zimmerli, 1984):

1. I describe the first relationship pattern between man, technology, nature, and culture as the "*judo*" *type*. By this I mean that technology, in the context of the recent sciences emerging from it, is understood as that sort of human craftsmanship which, by insight into and exploitation of natural laws, makes them serve man's own interests. In man's struggle with nature, he does not launch a frontal attack against the forces of nature; rather be attacks them calculatedly, employing the forces of nature themselves. In the same way that I defeat an opponent in judo, by vectorially adding the movement of his body and the force he expends to that applied by me, modern technology places itself "in the strength of the opponent." Francis Bacon, who has been erroneously

interpreted as the forefather of man's exploitative relationship with nature, can be taken as our chief witness in this connection. Nonetheless, until recently man has been taken to be the point of reference for the understanding of technology, and Bacon's philosophy is to blame. For Bacon, an exploitation of nature which goes against the personal interests of man would necessarily have been absurd and unimaginable. Man, using technology at the beginning of modern times, according to this profile, is *homo faber* – but only to the degree that nature allows him. We are dealing with *homo faber mensura naturae*.

2. With the Industrial Revolution, a second type of relationship emerged. I would call this the *"reproducibility/profit" type*. Prepared by a partial mechanization in manufacturing, technology had become definable by its utilization context; finished products emerge by means of technology. That the human workforce on a worldwide scale is being tendentiously and increasingly replaced due to the development and enforcement of mechanical and Taylorized production methods has been discussed widely. We can assume this as common knowledge. The qualitative jump to the new rests in the fact that products have now become technically *reproducible, and* technology thereby enters into a close relationship with a profit-oriented economy. This brings about not only an interchangeability of individual workers, but, as Karl Marx saw all too clearly (Marx, 1867 [1969, pp. 391 ff]). a general "de-skilling" and a reciprocal impact of production-technology on the social structure – on man's understanding of himself and others, including his culture. The natural rights view of modern liberals – the view, for instance, of John Locke who held that private ownership is an integral part of man's emergence from a "natural" condition – is now further developed. A new image of *homo faber* emerges, thriftily taking into account the balance sheets of his produce and hoping they will reproduce – an image of *homo faber oeconomicus*.

3. Another new configuration of the elements in this network of relationships takes place within the context of the "scientific and technological revolution" (STR); therein a third type emerges which we could call the *"white coat" type*. Science itself becomes a productive force, which means that even everyday technology is made scientific. This changes the image of technology considerably; the place of the mechanic, covered with grease and holding a wrench, is taken by the image of the

gentleman in a white coat who – as in a specialized garage, now called an "automotive diagnosis center" – reads off highly complicated instruments. For *homo faber*, this means that an original competence in controlling the technical understanding of the world and its changes has increasingly disappeared – replaced by *homo faber scientificus*. The non-technical person is capable of repairing and correcting the technical system he uses in only very few areas. Scientific technology and the mechanization of science have progressed to such an extent that *homo faber scientificus* ironically turns out to be *homo faber ignorans* with respect to the cognitive, technological, and social elements of his technical behavior.

4. With that we have reached our present time, the set of relationships I would label the *"awakening technological"* type. With the insight that, today, science and technology are mediated by information technology, something like a reflexive transformation of the technological/critical elements which were previously available emerges; the technocratic dream of correcting and mastering the world by mechanizing more and more spheres of life (cf. Veblen) is now recognized as a bad dream, and we are frightened awake by this nightmare to realize the perils of our current situation. This "reflexive transformation" (Zimmerli, 1978, pp. 124 ff and *passim*) occurs both within science and technology and in the public consciousness; attention is increasingly diverted to the unintentional effects or side effects which the mechanization of the world has brought about. The price that man is prepared to pay for additional comforts of a technological nature drops, and, consequently, other items rise in price – for example, the small remaining resources of a non-technological nature, non-mechanized agricultural systems, or a sphere of privacy that is not intruded upon by others. It is not at all by chance that this transformation parallels the "revolution of information technology" since the latter is intrinsic to the change in the technological age. The new pattern of thinking that stamps the new type of technology is cybernetically determined, a development which Arnold Gehlen (1957, p. 22) had already anticipated in the 1950s. However, the emphasis on technological innovation and optimization is, in this transformation, progressively shifted toward inventing and developing new forms of guidance technologies and other technologies to prevent side effects or to eliminate the undesired side effects of technology that has already been introduced and to restore nature to its original state. That

this constitutes, in the eyes of the public, a dispute between advocates and enemies of technology (Sieferle, 1984) is only an external manifestation of what is going on, is just one facet of it, and in any case is nothing new.

The consequences of this transformation for man's self-understanding are important. On the basis of a double inversion (MacCormac, 1986), man in the technological age sees himself as the technological analogue of his own products – that is, as a kind of data-processing computer (calculator). However, that is not the only consequence; the sense of insecurity in man that arises as a result of this reflexive transformation is very great. The "awakening" type of man in the technological age is very much torn apart and split. Not only is technology presented to him as being at the same time *both* desirable *and* abhorrent; moreover, he is dependent on technology to an extent never before known to him. He knows, however, that he has no chance of escaping from this dependence even though (as he likewise knows) it will in all probability lead to his downfall. And he has begun to sense that his natural basis, including his basic instincts – what typifies him as *homo faber* – can in principle be replaced technologically by way of the key technology of the 1990s, biotechnology applying the tools of genetic engineering. Thus he does not only not know what will become of him; he also does not know whether he ought to want these changes or not. The great progress made in knowledge has brought about a degree of ignorance that cannot be increased any further: *homo faber technologicus* becomes *homo faber doctus ignorans*.

I summarize: In spite of highly developed forms of technology which have developed in the modern and post-modern period, we still think of concepts and the theories arising from them in substantialist terms, and this leads to many unnecessary disputes. Instead of belligerent debates about the substance (nature, essence) of technology, it is more meaningful to interpret the problem area of man, technology, nature, and culture as a network of relationships linking various nodal points, to reconstruct the changes undergone in various stages. This reconstruction seems to reveal four stages. *Homo faber mensura naturae* corresponds to the early modern "outsmarting" type of technology and to a concept of nature regarded as the outsmarted opponent. *Homo faber oeconomicus*, with nature recognized as being the other thing necessary for man's self-disposal and self-realization, corresponds to the "reproducibility-profit" type. *Homo faber scientificus*, to whom nature and culture constitute separate areas, corresponds to the "white coat" type of the scientific and technological revolution. Finally, *homo faber doctus ignorans*, whose relationship to nature and the world is completely severed and torn to pieces, corresponds to the "reflective" type of the awakening stage. The relationship between man, technology, and nature – which, as a network, constitutes culture – can in the "reflective" type stage reveal, with reference to the basic structure uniting these elements, that all the stages are defined by their specific relationship of tool-using man to technology and nature.

II

If, with regard to technology, the question we want to raise is that of responsibility, then the concept of responsibility concerning developments up to our day should be analyzed on a non-substantialist basis (just as with the concepts of "technology," "man," "nature," and "culture"). The context we are dealing with is that of a network of relationships, the nodal points of which constitute only apparently fixed concepts. For this reason, I formulate a second thesis.

(II) *The concept of responsibility means, in the first instance, no more than the linguistic reflection of the fact of being related. In other words, I can only be held responsible for those activities which I have, to one degree or another, brought about. It would never occur to anyone to hold someone responsible for the sun's rising merely because it happened to coincide with something that other person is doing. If we suppose that in the end everything is connected with everything else, then everyone is more or less directly responsible for everything that happens. However, the concept of responsibility in the narrower sense refers above all and only to those actions which were triggered or partly triggered off by the person in question. The subject must answer for actions he has caused or partially caused. He becomes thereby the subject or responsibility.*

"Responsibility" is thus the reflexive name for the formal character of dependence among beings. This formal dependence has various instantiations which must be gradually distinguished. This is mentioned in order to point out that in every single case the term "responsibility" must be defined afresh. That is, if what has been said is true and the term "responsibility" reflects only the *formal* aspect of dependence, each case must be materially different. Therefore, the *content* of the concept of "responsibility" must remain historically contingent. (More will be said about this later.)

Moreover, dependence is not the only aspect of contingency here. We must also consider the historically different forms of *justification* with regard to moral demands; these too enter into the concept of responsibility. As a medieval knight, for instance, it may have been my responsibility not to put up with affronts that would imply disgrace or shame; I would have to challenge my assailant to a duel. Today the thought of a duel to protect one's "honor" seems ridiculous. Thus we can expect a different concept of responsibility in each of the four stages of modern technology.

Something else emerges from these considerations, which I would like to formulate as follows in a third thesis:

(III) *The popular contemporary form of responsibility connected with technology and the natural sciences takes its primary orientation from late medieval concepts of responsibility associated with the craft guilds. In large part, the difficulties we encounter in our contemporary situation with the technology that dominates it arise from the lack of congruity between technological development and this sort of responsibility* (Zimmerli, 1985).

The understanding of responsibility of the craftsmen's guilds – which has remained almost unchanged since the beginning of the modern era – involved, on one hand, the responsibility of the craftsman/mechanic toward his customer for the quality of his product and the correctness of their business relationship. For example, it was the responsibility of the gunsmith that the barrel he had forged should give the bullet the right momentum so that it traces as long a trajectory as possible. On the other hand, responsibility was to a large extent also understood as an internal relationship with the members of the guild and with the guild as a whole. The "magical unity of the symbolic world," largely preserved in the traditions of work and tools, sees to it that dishonorable behavior and infringements of regulations are punished by actions that take on something of the character of a "magical cleansing ritual" (Sieferle, 1984, p. 67). A kind of self-control of the craftsman's practice results, as expressed in the charter of Thorn of 1523: "No craftsman is to devise, invent, or use anything new, but is to follow his neighbor out of civil and brotherly love" (cf. Rüstow, 1951, p. 380).

I call these two forms of responsibility "internal." They are (a) the internal responsibility toward the customer for the quality of the product and the correctness of their business relationship, and (b) the internal responsibility of the craftsmen toward the group and its members for the upholding of standards within the group and for preventing the lowering of these standards.

If, on this basis, we now take a close look at the developing patterns of the man–technology–nature–culture relationships which I have suggested, we see that these characteristics of guild responsibility can be applied appropriately only to phases 1 and 2. Already in phase 3 it becomes questionable whether this interpretation suffices for this differentiated form of technology. The reducing of human responsibility to these two kinds of internal responsibility is undoubtedly not sufficient; the "scientific technological type" of phase 3 has become more and more highly specialized, with the result that the consequences of particular aspects of this form of technology can no longer be appraised separately.

What is the problem here? Well, it is obvious that the idea of technologists and natural scientists bearing only internal responsibility has become obsolete in cases of technology that provoke a large number of secondary and unintended consequences. We must now assume a whole series of *external* responsibility relationships which are not related to the production and distribution process alone. And the whole matter becomes extremely complex if it is a matter of responsibility relationships concerning certain circumstances which we have only partly brought about ourselves, which we can not properly be said to have caused – and so be said to be guilty for.

In order to do justice to this, it is essential to analyze carefully both the concept of responsibility and its connection with actions and consequences.

First of all, in order to get a grip on what still remains unclear, I would like to analyze and define the concept of "being responsible" in a fourth thesis:

(IV) *"Being responsible" is a three-termed relation involving someone (the subject of responsibility) who is responsible to someone or something else (the forum of responsibility) for someone or something (the scope of responsibility).*

We should bear in mind the medieval source of this terminology: "being responsible" refers to a situation in which the *person* in question (subject of responsibility) must give an account of his *actions* (scope of responsibility) before the *court* (forum of responsibility). In this context, only those people who are of age – that is, those who can speak for themselves and who require no guardian – can be responsible.

For medieval man it was clear that when it came down to it, the ultimate forum of responsibility was God's judgment; the subject of responsibility was himself; and the scope of responsibility was what, earlier, I called internal responsibility. But since the beginning of the modern era, uncertainty has set in; it was already problematic when the Christian faith was still widely shared, for it is not known whether this faith suffices to form a basis for moral norms, or whether reason and not faith had already then become the decisive factor. The last three hundred years have been characterized by attempts to develop ethics in a non-religious way. From Descartes' "provisional ethics," *via* Kant's categorical imperative, to the utilitarian approach of the nineteenth century, we witness the most varied attempts to solve this problem. During the course of this development, and in accord with the basic ideology of the time, the opinion on what is to be regarded as the ultimate forum of responsibility shifted. Our contemporary situation reflects an expansion of the notion of a forum of responsibility a degree not known previously. From the beginning, the modern age has been characterized as "secularized." That has always meant uncertainty about the forum of responsibility. The human ego first takes the place that God had held; then, in the Enlightenment, the entirety of all living, adult men (that is, men capable of bearing responsibility) becomes the forum; and finally, of late, future generations ("intergenerational justice") and extra-human nature are added to the list.

Parallel to this, there is also an expansion of the *scope* of responsibility. Each new discovery and each social implementation of some new technology in industrial production leads to an increase in possibilities of action – and therefore increases in responsibility. And this remains so even if responsibility is still interpreted as purely internal (Lenk, 1979, pp. 69 ff; 1982, pp. 198 ff).

A final and most significant change is closely related with this, namely, in the *subject* of responsibility. That the actions of any one individual render him the subject of responsibility pervades both traditional and modern ethics. However, this view becomes increasingly questionable in the sphere of technological action. No single technological action results from an individual's will to act. Many persons play a direct or indirect role in any decision an individual makes, and the real subject of action is always a collective, a team or a group. Added to this is the fact that the results of actions or of the application of new forms of key technologies – e.g., information technology, or robotics, or biotechnology – are basically never totally foreseeable. (I have called this the "paradox of information technology," Zimmerli, 1986, pp. 296 ff.) This was also true to a certain extent for every earlier form of technology, but is valid here in a qualitatively new way. Whereas previously at least the capacities of controlling and directing were left up to man – leaving him thereby responsible – even these functions are apt today to be fulfilled by machines in the branches of information technology and robotics; and in the branch of biotechnology they are left to organisms which have been changed by means of gene technology.

I summarize: An appropriate change in the concept of responsibility, corresponding to the differing typologies of the man–technology–nature–culture network, has not taken place. Rather, even today, as far as technology and nature are concerned, the concept of responsibility retains the pattern of the responsibility ideal of the medieval guilds of craftsmen, and that pattern involved purely internal responsibility – toward the client for his product, and toward his guild colleagues and the guild as a whole. The producer was liable to the client for the quality of his product, and for the correctness of their business relationship; he was responsible to the guild for the maintaining of its standards. The guild craftsman was not liable for the use or possible misuse of his product. Furthermore, responsibility, on careful analysis, turns out to be a three-term relationship linking the *subject*, *scope*, and *forum* of responsibility. But all three of these elements have changed decisively in the course of secularization in the modern period. The entirety of rational beings, present and future, and, should the occasion arise, extra-human nature, replaces God as the forum of responsibility. The scope of responsibility is expanded to encompass all new forms of technology, even those of which man is aware that the consequences are in principle unforeseeable. This is linked with the most fundamental change, in the subject of responsibility: man must now, quite obviously, admit both his limitations as an individual, and his limitations even in actions which he himself consciously steers and for which he would be held responsible.

III

If we bear in mind, in accord with this analysis, the divergence of types of technology from types of responsibility, it becomes clear that we are at a point where something qualitatively new is happening. However, the situation cannot be accounted for exclusively on the basis of history. If, as with the traditional definition, "being responsible" is understood as a three-term relationship, then an agent (subject of responsibility) must answer for that which his action has brought about (scope of responsibility) to the court (forum of responsibility).

Because of the novelty of the conditions sketched, this idea must be formulated in a novel way: namely, if we look more closely, we see that in the traditional conception the subject of an action and the subject of responsibility merge with one another. We can theoretically analyze and determine the conscious subject of an action as so and so, because he knows that a certain action a_1 gives rise to consequences $c_1 - c_n$, and because he would like at least to effect the consequences $c_1 - c_4$, he performs the action a_1 and also willingly accepts the consequences $c_5 - c_n$, for which he is then also responsible. On the other hand, still in exact accord with the same classical view, in the case of an acting individual knowing only the consequences $c_1 - c_4$ of an action a_1, he can only be held morally responsible for the consequences known to him, and not for those unknown to him. (It is a different matter in the case of legal liability, where every consequence of a certain action may be ascribed to its subject.) Now the novelty of the current situation can be described thus. On the one hand, we can be held responsible even for actions of which we are not the subject, and, on the other hand, moral responsibility remains even in cases in which we have not reckoned with the consequences or where we were *unable* to reckon with the consequences due to the specific nature of the new forms of technology.

In other words, when in the technological age – i.e., in the reflection-type stage of technology – the human individual, as the *homo faber doctus ignorans*, is not only no longer identical with the subject of an action but he also *knows* that the subject of action is constituted by teams, groups, collectives, and large concerns, the subject of action and the subject of responsibility diverge because, at the same time the reflexive form of technological knowledge requires that, by definition, I can never assess the consequences of my actions, it also requires a subject capable of moral responsibility. I therefore formulate a fifth thesis:

(V) *Even when the subject of an action shifts more and more toward teams, groups, and collectives, the subject of responsibility does* not *shift: but it remains the* individual.

In this, the old internal responsibility – for product and to guild – cannot be removed; this is self-evident. But in this stage another external form is added to responsibility in the sense of liability for possible unintended uses of the product in question. This means that the risk of an unforeseeable use of the product must be included when considering the scope of responsibility.

Something similar applies to the *forum* of responsibility. Here also the subject of responsibility becomes evident only when the subject performing an action and the subject of responsibility are separated. The reciprocal effect is a differentiation among forums of responsibility, to which the technological producer must answer. In addition to the forum of the client (which, as before, coincides with an internal responsibility relationship), there now exists also the abstract external forum of all rational beings – which means the public. And in addition to the group forum of internal professional responsibility (which has replaced that of the guild), there exists a new forum – all those who could potentially be harmed by any unintended side effects. These persons (and things) function as it were as "accessory prosecutors."

This formulation shows the difference between "responsibility" and "liability," but it is necessary to say something more about the difference between moral and legal responsibility. With the reflexive turning point in the history of technology, the difference seems to grow less and less sharp; it almost seems as if law and morality, responsibility and liability have coalesced. And yet we must hold on to a decisive difference – as should be apparent from what has been said. If in fact, as I said at the beginning, responsibility is nothing more than the conscious, abstract form of a relationship of mutual dependence, it follows that man's responsibility takes concrete forms in successive phases, with each instantiating new responsibilities for everything and to everyone with whom man enters into new relationships. It follows further that, were it not for the fact that legal liability lacks subjective awareness, it too could be thought of as an instance of responsibility. Or, put another way, the subjective feeling-oneself-responsible does not always coincide exactly with the objective relationship of being responsible. The latter, as Nietzsche emphasized, is a product of upbringing; but instead of drawing nihilistic conclusions from this, it would be far more meaningful to draw a positive conclusion. One of the tasks of the future is to transform objective relationships of responsibility into a subjectively felt sense of responsibility. What we should call our "sense of responsibility" – even though it is ultimately reduced to the medieval model of the craftmen's guilds (as far as the problematic of man, technology, nature, and culture is concerned) – should be the product of a new sort of upbringing which strengthens the sense of responsibility in contexts where, previously, only objective liability was thought to exist. This would be a strong conclusion to be drawn from my analysis. Ethical reflection – which in the context of modern rationalist thinking always means formal ethical reflection – is not thereby tied to the letter of the law. On the contrary, only in this way does ethics have the chance to develop beyond the limits of legal requirements. Only if the relationship between moral deliberations and legal requirements is shifted in a positive direction, with the new relationship viewed as advantageous, can we readily accept it.

When we consider this situation more closely, we see clearly that neither the relationship between law and morality nor the structure of responsibility itself has changed: individuals are still responsible for the consequences of the actions they perform. All that has changed is the addition of something new to the internal concept of responsibility and the coalescing of the subject of an action and the subject of responsibility. In order that, in the long run, dysfunctional and cognitive discord does not result from this, the continued awareness of a difference between legality and morality must be balanced: our sense of responsibility for that which we have not ourselves, or not alone, caused, grows by practicing the concept of responsibility against a background of the difference between the subject of an action and the subject of responsibility.

I sum up: A theoretical analysis shows that it only seems to be the case that the subject of an action and the subject of responsibility always coincide. In the new forms of technology (and the reflective ideal type they characterize), we can see clearly that the subject of action and the subject of responsibility are different. As a consequence, the subject of responsibility must even answer for actions which were not carried out or caused by him alone. However, and conversely, this also has the effect of fragmenting the medieval forum of responsibility to the client legitimized by God, turning it into a variety of partial forums. An examination of this shift, furthermore, requires that we consider the connection between moral and legal responsibility, for the objective relation of legal liability, in this stage of technological development, clearly contrasts with the subjective relation or awareness or feeling of responsibility. This leads to a final conclusion, that in the new era education bears the burden of training us in the practice of the new external sense of responsibility in place of the old internal sense. We are obliged to promote the emergence of a new sense of wider responsibility.

Technical University Carolo-Wilhelmina, Braunschweig

REFERENCES

Gehlen, Arnold. *Die Seele im technischen Zeitalter* (Hamburg, 1957).

Hart, H. L. A. *Punishment and Responsibility* (Oxford, 1968).

Jonas, Hans. *The Imperative of Responsibility: In Search of an Ethics for the Technological Age* (Chicago, 1984; original German, 1979).

Lenk, Hans. *Pragmatische Vernunft* (Stuttgart, 1979).

Lenk, Hans. *Sozialphilosophie der Technik* (Frankfurt, 1982).

Lenk, Hans. 'Zum Verantwortungsproblem in Wissenschaft und Technik', in E. Ströker, ed., *Ethik der Wissenschaften?* (see below).

MacCormac, Earl. 'Men and Machines: The Computational Metaphor', in C. Mitcham and A. Huning, eds., *Philosophy and Technology II: Information Technology and Computers in Theory and Practice* (Dordrecht, 1986).

Marx, Karl. *Das Kapital*, vol. 1 (Berlin, 1969; original, 1867).

Neumeier, O., ed. *Wissen und Gewissen: Arbeiten zur Verantwortungsproblematik* (Vienna, 1986).

Radnitzky, Gerard. 'Responsibility in Science and in the Decisions about the Use or Non-Use of Technologies', in O. Neumeier, ed., *Wissen und Gewissen* (above).

Rüstow, A. 'Kritik des technischen Fortschritts', *Ordo* 4 (1951): 373–407.

Sieferle, R. P. *Fortschrittsfeinde? Opposition gegen Technik und Industrie von der Romantik bis zur Gegenwart* (Munich, 1984).

Ströker, Elisabeth, ed. *Ethik der Wissenschaften?* (Munich, Paderborn, and Vienna, 1984).

Weber, Max. 'Politik als Beruf' (1919), reprinted with the title, 'Der Beruf der Politik', in

M. Weber, *Soziologie: Weltgeschichtliche Analysen: Politik*, ed. J. Winckelmann (Stuttgart, 1964).

Zimmerli, W. Ch. 'Von Kernenergie und menschlicher Verantwortung am Ende des Wachstumsdenkens', in W. Zimmerli, ed., *Kernenergie – wozu?* (Basel and Stuttgart, 1978).

Zimmerli, W. Ch. 'Mut zur Furcht: Facetten technischer Humanität in Vergangenheit und Zukunft', *Mitteilungen der Technischen Universität Carolo-Wilhelmina zu Braunschweig* (1984): 33–40.

Zimmerli, W. Ch. 'Jenseits der individuellen Verantwortung: Rüstung und Ethik im technologischen Zeitalter', in H. Bähren and J. Tatz, eds., *Wissenschaft und Rüstung* (Braunschweig, 1985).

Zimmerli, W. Ch. 'Who Is To Blame for Data Pollution? On Individual Moral Responsibility with Information Technology', in C. Mitcham and A. Huning, eds., *Philosophy and Technology II* (see above).

NOTE

A German version of this paper will appear in H. Lenk and G. Ropohl, eds., *Technik und Ethik* (Stuttgart: Reclam, 1987).

EDMUND F. BYRNE

WORK AND TECHNOLOGY: A BIBLIOGRAPHICAL ESSAY

A complete library of writings on the interface between work and technology would include everything from Aristotle's reflections in the *Nichomachean Ethics* about machine replacements for slaves up to the latest rationale for automation. The task of producing a definitive catalogue for that library would be not Herculean but Procrustean in view of the fluid definitions of "work" and of "technology," the many related topics that touch on either, and the wide variety of specialists who have addressed them. But only a subset of these works contain or invite deliberations with philosophical import; and fewer still are authored by professional philosophers. Such qualifications notwithstanding, there does exist a literature that can be subsumed under the title: Work and Technology in Philosophical Perspective.

Any attempt to neatly categorize such literature would be more pedantic than illuminating at this stage. But it is possible to suggest within a continuum three different areas of emphasis with which a particular writing might be identified, namely, philosophical deliberation with regard to: (1) the incorporation of technology into work processes; (2) the comparative value of work and technology respectively; and (3) the appropriate "division of labor" between work and technology in the future.

These areas are not mutually exclusive; each inevitably points to the others. How they differ can be suggested by a kind of stylized chronology. On the assumption that a society's values with regard to work are fairly well established, the introduction of a work-related technology typically generates philosophical reflection in direct proportion to the perceived gravity of its consequences. On the first level of philosophical reflection are addressed technical and ethical questions about the actual or eventual impact of technology on work. Once this impact is reasonably well understood, more abstract questions arise about the comparative value of work and technology, respectively, and about the values that underly social commitment to either of these. Such questions having been adequately addressed, one is then faced with a policy-oriented, and potentially utopian, question: what should be the relationship

Paul T. Durbin (ed.), Technology and Contemporary Life, 295–313.
© 1988 by Edmund F. Byrne.

between work and technology in the future? A philosophical consideration of the relationship between technology and work will inevitably
touch on all three areas; but it might emphasize only one or two of
them.

1. PUTTING TECHNOLOGY TO WORK

The impact of technology on work may be studied with an emphasis on
(a) the work impacted upon; or (b) the technology having the impact.

A. *Work as Affected by Technology*

The most basic philosophical claim about the impact of technology on
work is that tool-making is somehow the evolutionary source of those
species-specific traits that make human work possible. But tool-making
is itself labor; so why not recognize labor as the origin of culture?
Known as the **labor theory of culture**, this very claim was suggested by
Karl Marx and developed by Friedrich Engels. Charles Woolfson, a
sociologist, reviews the evidence in his *The Labour Theory of Culture: A
Re-examination of Engels's Theory of Human Origins* (London: Routledge & Kegan Paul, 1982). See also Clifford Geertz, 'The Impact of the
Concept of Culture on the Concept of Man', in *New Views of the Nature
of Man*, ed. J. R. Platt (Chicago and London: University of Chicago
Press, 1965), pp. 93–118. Singularly important as a philosophical refinement of this originary hypothesis is Georg Lukács's study of "the
teleology of labour," published posthumously as *The Ontology of Social
Being: 3. Labour* (trans. David Fernbach; London: Merlin Press, 1980).
    In a broad sense any organization of a workforce constitutes a social
technology that affects work in various ways. The systematic utilization
of slave labor to construct a pyramid or to propel a galley ship are
familiar examples; as are the bureaucracies that sociologists from Weber
to Ellul have analyzed. These bureaucracies, viewed as structured
career ladders, are the subject of *Employing Bureaucracy: Managers,
Unions, and the Transformation of Work in American Industry, 1900–
1945*, ed. Sanford M. Jacoby (New York: Columbia University Press,
1986).
    The implications of such human-based technologies have been stud-
ied philosophically to some extent. In a section of the *Grundrisse*
recently published as *Pre-Capitalist Economic Formations* (trans. Jack

Cohen; London: Lawrence & Wishart, 1964), Karl Marx proposed a dynamic theory of pre-capitalist economic formations. That theory is brought up to date against the background of subsequent research by Barry Hindess and Paul Q. Hirst, *Pre-Capitalist Modes of Production* (London: Routledge & Kegan Paul, 1975).

Related to these considerations are all the utopian writings that have proposed an allegedly better way to organize people to accomplish the work that needs to be done in a way most congenial to the people involved. An excellent introduction to this literature is Frank E. Manuel and Fritzie P. Manuel's 896-page study, *Utopian Thought in the Western World* (Cambridge, MA: Harvard/Belknap, 1979), which includes a helpful bibliography. More focused on the question of organizing work to the benefit of the community is J. C. Davis, *Utopia and the Ideal Society: A Study of English Utopian Writing, 1516–1700* (Cambridge: Cambridge University Press, 1981); and, for example, Stanley Buder's account of a company town, *Pullman: An Experiment in Industrial Order and Community Planning, 1880–1930* (Oxford: Oxford University Press, 1967).

More overtly concerned with methodology are those who dream of getting maximum productivity out of the workforce through some version of engineering. This dream had already been recorded in such works as Charles Babbages' *On the Economy of Machinery and Manufactures* (1832): Frederick Winslow Taylor's *Scientific Management* (1911); and, with a focus on management. Thorstein Veblen's *The Theory of the Leisure Class* (1899, 1912) and especially his *The Engineers and the Price System* (1923).

The classical economists fostered just this way of viewing the impact of technology on work with their respective versions of a **labor theory of value**. The issues explored in that context are carefully articulated by Ronald L. Meek, *Studies in the Labor Theory of Value* (2nd ed.; New York/London: Monthly Review Press, 1965). In Volume I of *Capital* (trans. Ben Fawkes; New York: Random House, 1977), Karl Marx supports his critique of capitalism by means of what may be the first detailed analysis of the impact of new technologies on working class individuals and families. Subsequent developments are insightfully evaluated from the natural law perspective by Yves R. Simon, *Work, Society and Culture*, ed. Vukan Kuic (New York: Fordham University Press, 1971).

The long ignored economic implications of Marx's labor theory of value have in recent years been reintroduced into Western economics by means of linear reproduction models of a capitalist economy that take into account class conflict as well as efficiency and mutual satisfaction. See Robert Paul Wolff's *Understanding Marx* (Princeton, NJ: Princeton University Press, 1984). Analysis of labor as a market phenomenon is, of course, a standard concern of economists, some of whom even analyze worker ownership. See Ugo Pagano, *Work and Welfare in Economic Theory* (New York/Oxford: Basil Blackwell, 1985); and Norman J. Ireland and Peter J. Law, *The Economics of Labour-Managed Enterprises* (London and Canberra: Croom Helm, 1982), which includes a helpful bibliography. But the economics of labor has not been of great interest to philosophers. It is barely mentioned, for example, in Daniel M. Hausman, ed., *The Philosophy of Economics: An Anthology* (Cambridge: Cambridge University Press, 1984.)

Marx's theory of **alienation** has generated a great deal of theorizing among sociologists and philosophers but far less among Marxists who associate the notion with the early ("humanist") Marx. See Erich Fromm's introduction to the early manuscripts in *Marx's Concept of Man* (New York: Ungar, 1961). In recent years some promising and creative theorizing about work centers around the notion of alienation and its close cousins, estrangement and reification. Kostas Axelos focuses on Marx in *Alienation, Praxis and Techne in the Thought of Karl Marx* (trans. Ronald Bruzina; Austin and London: University of Texas Press, 1976). The Marxist literature is reviewed by Richard Schacht in his *Alienation* (New York: Doubleday, 1971). Two international, cross-disciplinary conferences resulted in proceedings edited by R. Felix Geyer and David Schweitzer: *Theories of Alienation: Critical Perspectives in Philosophy and the Social Sciences* (The Hague: Martinus Nijhoff, 1976), and *Alienation: Problems of Meaning, Theory and Method* (London/Boston/Henley: Routledge & Kegan Paul, 1981).

Of philosophical significance in the latter work are: Peter Christian Ludz's review of the *positive* meanings of alienation in the history of philosophy; Richard Schacht's distinction between a descriptive and an evaluative concept of alienation; John Torrance's preference for a structural sociological theory detached from "alienation"; John Horton and Manuel Moreno's assessment of the role of alienation in the thought of Althusser and Braverman; Mihailo Markovic's insistence upon the need for "self-determination" beyond workers' participation, control, or even self-management; and Walter R. Heinz's argument that "preoccupational socialization prepares the individual for a normative acceptance of the conditions of alienated work."

Of related import are two articles in a volume of *The Philosophical Forum* **10**, nos. 2–4 (Winter–Summer, 1978–1979) devoted to the topic of work: Frederick M. Gordon, 'Marx's Concept of Alienation and Empirical Sociological Research' (pp. 242–264); and Judith Buber Agassi, 'Alienation from Work: A Conceptual Analysis' (pp. 265–305; includes responses), which discusses the theoretical dispute among Marxists over the revolutionary potential of job dissatisfaction. In disagreement with André Gorz, Joachim Israel insists that in the absence of class consciousness, job dissatisfaction will never lead to revolution: *Alienation: From Marx to Modern Sociology: A Macro-Sociological Analysis* (Boston: Allyn and Bacon, 1971).

Directly interrelating alienation and technology is Simon Marcson's *Automation, Alienation and Anomie* (New York: Harper & Row, 1970) and Jon M. Shepard, *Automation and Alienation: A Study of Office and Factory Workers* (Cambridge, MA: MIT Press, 1971). Related to these studies is the work of social historians and others who, ironically, attribute the absence of a working class consciousness in the United States to the "homogenizing" effect of various new technologies. See David M. Gordon, Richard Edwards, and Michael Reich, *Segmented Work, Divided Workers: The Historical Transformation of Labor in the United States* (Cambridge: Cambridge University Press, 1982). For alternative perspectives, see David Montgomery, *Workers' Control in America* (Cambridge: Cambridge University Press, 1979); Wolfgang Abendroth, *A Short History of the European Working Class* (New York and London: Monthly Review Press, 1972); and Alejandro Portes and John Walton, *Labor, Class, and the International System* (New York: Academic Press, 1981), which includes an extensive bibliography.

In contrast to Engels and his successors, Lewis Mumford looks to play rather than to work for the etiology of thinking, e.g., in *The Myth of the Machine: Technics and Human Development* (New York: Harcourt Brace Jovanovich, 1967). In this respect he justifies the preference of (bourgeois) philosophers for studying not the process of work but that of thought. In particular, philosophers (and others) have for centuries been exploring the **technology of thinking**. Ramon Lull, Leibniz, Pascal, Peirce, Turing and others have speculated about the possibility of building a machine that could accomplish at least some of the feats we attribute to human thinking. This speculation, today being concentrated on AI research, is encouraged by the thesis of J. O. de La Mettrie, *L'Homme Machine: A Study in the Origins of an Idea* (critical edition; Princeton NJ: Princeton University Press, 1960).

Entry into this literature is facilitated by the bibliographical information provided in Carl Mitcham and Alois Huning, eds., *Philosophy and Technology II: Information Technology and Computers in Theory and Practice* (Dordrecht: D. Reidel, 1986), pp. 307–339. See in particular Irene Taviss, *Technology and Work* (Cambridge, MA: Harvard University Press; originally published as *Research Review No. 2*, Winter, 1969); Alfred Chapuis and Edmund Droz, *Automata: A Historical and Technological Study* (trans. Alec Reid; Neuchatel: Editions du Griffon, 1958); and Egmont Hiller, *Automaten und Menschen* (Stuttgart: Deutsche Verlag-Anstalt, 1958).

In general, analyses by philosophers tend to be more skeptical than are those of technical experts in the "brain machine" industry about the level of competence that such machines might attain. See in this regard the Stanford Research Institute, *Management Decisions to Automate* (Project No. ISU–4530, Menlo Park, CA: SRI, 1964), and the Carnegie–Mellon University study, cited below. Joseph P. Engelberger, a robot manufacturer, touts the labor saving implications of his wares in *Robotics in Practice* (New York: American Management Association, 1980); Engelberger's heritage may also be learned from Henry Elsner, *The Technocrats: Prophets of Automation* (Syracuse, NY: Syracuse University Press, 1967).

B. *The Extent of Technology's Impact on Workers*

Whatever the likelihood of a perfect brain machine, we already need to ask what impact even imperfect thinking machines have on human work and workers. The impact may be considered positive in many respects; but it is also likely to be negative. Whence the ambivalence of Robert Hugh Macmillan's work entitled *Automation, Friend or Foe?* (Cambridge: Cambridge University Press, 1956). Just as ambivalent is a 1984 study by the U.S. Office of Technology Assessment, *Computerized Manufacturing Automation: Employment, Education and the Workplace*, which distinguishes and evaluates four basic strategies the government might adopt: laissez faire (identified with current policies); (2) technology-oriented, i.e., emphasis on programmable automation development and use; and (3) resource-oriented, i.e., "upfront attention to education and the work environment, and job creation"; or (4) a combination of items (3) and (4). Ongoing analysis of these and other related issues will be the function of a promising new semiannual publication entitled *Technology, Work and Employment*, edited by Colin Gill and published by Basil Blackwell.

H. J. Habakkuk's *American and British Technology in the Nineteenth Century: The Search for Labor Saving Inventions* (Cambridge: Cambridge University Press, 1962) lent credence a century later to the validity of Marx's thoughts about technology's impact on workers. But at the time many others foresaw only the benefits of the coming automation: e.g., Leon Bagrit, *The Age of Automation* (New York: Mentor, 1965); or, like George Terborgh in *Automation Hysteria* (New York: Norton, 1966), they chastised those who expected a more negative impact. More recently, automation is held blameless for any future unemployment in a management-sponsored study by H. Allan Hunt and Timothy L. Hunt, *Human Resource Implications of Robotics* (Kalamazoo, MI: W. E. Upjohn Institute for Employment Research, 1983), and the problem is found only mildly worrisome in a Carnegie-Mellon University study entitled *The Impacts of Robotics on the Workforce and Workplace* (Pittsburgh, 1981).

Today few responsible analysts of this process deny that worker displacement is an inevitable *short-term* consequence of introducing labor saving technology. See, for example, Colin Gill, *Work, Unemployment and the New Technology* (London: Basil Blackwell, 1985). But in general social scientists other than those with Marxist leanings have been rather blasé about the *long-term* impact of technology on work. Typical in this respect are the writings of Herbert A. Simon and John Diebold. Daniel Bell practically eliminates industrial work in his vision of a society run by science-sensitive professionals, *The Coming of Post-Industrial Society* (New York: Basic Books, 1973); but he ignores the worker displacement that scenario implies. However, the labor process claim (see below) is carefully evaluated against the background of sociological theories about work by Stephen Hill, *Competition and Control at Work: The New Industrial Sociology* (Cambridge, MA: MIT Press, 1981). And political sociologist Claus Offe answers affirmatively his own question: 'Work: The Key Sociological Category?' in *Disorganized Capitalism: Contemporary Transformations of Work and Politics* (Cambridge, MA: MIT Press, 1985), pp. 129–150.

The range of possible impacts of technology on work can be gleaned from Annette Harrison, *Bibliography on Automation and Technological Change and Studies of the Future* (Rand Paper P–3365–3; Santa Monica, CA: Rand, 1967; Springfield, VA: Clearinghouse for Federal Scientific and Technical Information, U.S. Dept. of Commerce, 1968); and RO-BOMATICS, a state-of-the-art microfiche service available in technological libraries and accessible *via* indexes and abstracts.

More alert to negative consequences are defenders of the labor movement, among whom must be included such humane geniuses as Norbert Wiener. Especially in his *The Human Use of Human Beings* (Boston: Houghton Mifflin, 1950) and *God & Golem, Inc.* (Cambridge, MA: MIT Press, 1964), Wiener joins a long tradition of writers who have seen clouds on the horizon because of the unregulated introduction of new technology into the workplace: e.g., Georges Friedmann, *Industrial Society: The Emergence of the Human Problems of Automation* (trans. Harold L. Sheppard; Glencoe: Free Press, 1955); Stuart Chase, *Men and Machines* (New York: Macmillan, 1943); Elliott Dunlap Smith, *Technology and Labor: A Study of the Human Problems of Labour Saving* (London: Oxford University Press, 1939); George E. Barnett, *Chapters on Machinery and Labor* (1926, reprinted 1969 by Southern Illinois University and Feffer & Simons), which focuses on the printing, stonecutting and bottling industries; and, in the preceding century (in addition to Marx), John Cameron Simonds and John T. McEnnis, *The Story of Manual Labor* (Chicago: R. S. Peale & Co., 1887); and Henry George's *Progress and Poverty* (1879). More recent warnings are those of Ben B. Seligman, *Most Notorious Victory: Man in an Age of Automation* (New York: Free Press, 1966); Robert Howard, *Brave New Workplace* (New York: Viking, 1985), which considers both workplace health and safety and worker displacement; Joan M. Greenbaum, *In the Name of Efficiency: Management Theory and Shopfloor Practice in Data-Processing Work* (Philadelphia: Temple University Press, 1979); and Harley Shaiken, *Work Transformed: Automation & Labor in the Computer Age* (Lexington, MA: D.C. Heath, 1986).

Typical of union responses to the displacement of workers by technology are the following: John Evans, *Negotiating Technological Change* (Brussels: European Trade Union Institute, 1982), and two publications of the AFL–CIO Department for Professional Employees (both published Washington, DC, 1981): Kevin Murphy, *Technological Change Clauses in Collective Bargaining Agreements*; Dennis Chamot and Michael D. Dymmel, *Cooperation or Conflict: European Experiences with*

Technological Change at the Workplace. See also Benjamin Sollow Kirsh, *Automation and Collective Bargaining* (New York: Central Book, 1964), and, for purposes of comparison, Milton J. Nadworny, *Scientific Management and the Unions, 1900–1932* (Cambridge, MA: Harvard University Press, 1955). The respective approaches of labor, management, and government are studied in *Technological Change: The Tripartite Response* (Washington, DC: International Labor Office, 1985).

The impact of labor saving technology on women in particular is being studied today as it was by Marx and others a century ago. See, for example, Heather Menzies, *Women and the Chip: Case Studies of the Effects of Informatics on Employment in Canada* (Montreal: Institute for Research on Public Policy, 1981), and two studies of work in the home: Ruth Schwartz Cowan in *More Work for Mother* (New York: Basic Books, 1983) and Susan Strasser, *Never Done* (New York: Pantheon, 1982). These and many other related studies Joan Rothschild incorporates into a groundbreaking anthology, *Machina ex Dea: Feminist Perspectives on Technology* (New York: Pergamon, 1983).

Somewhat more polemical is research that focuses on what is called the **labor process**. In Marxist parlance this has to do with the thesis that the ruling class introduces technology for the explicit purpose of controlling and when possible dispensing with workers. Seminal for recent research along these lines is Harry Braverman's *Labor and Monopoly Capital* (New York/London: Monthly Review Press, 1974), which argues that management decisions to introduce new technology have resulted in the "deskilling" of the workforce, that this result is intentional on the part of management, and that it is now reaching beyond the assembly line into the office. Follow-up studies include these works: Andrew Zimbalist, ed., *Case Studies on the Labor Process* (New York/London: Monthly Review Press, 1979), and Les Levidow and Bob Young, eds., *Science, Technology and the Labour Process: Marxist Studies,* 2 vols. (London: Free Association Books, and Atlantic Highlands, NJ: Humanities Press, 1981 and 1985). Historical studies in this perspective include Dan Clawson, *Bureaucracy and the Labor Process: The Transformation of U.S. Industry, 1860–1920* (New York and London: Monthly Review Press, 1980), and David F. Noble's studies, of the engineering profession in *America by Design* (New York: Knopf, 1977) and of numerical machine control in *Forces of Production* (Knopf, 1984). Challenging the explanatory potential of this thesis of the managerial quest for control (rather than for profit) are articles in Stephen Wood, ed., *The Degradation of Work? Skill, Deskilling and the Labour Process* (London: Hutchinson, 1982), which includes a bibliography.

Variations on the labor process theme are found in the writings of a number of philosophers. Anthony Skillen writes out of this perspective in 'The Politics of Production', a chapter of *Ruling Illusions* (Atlantic Highlands, NJ: Humanities Press, 1978). Mark Okrent, in 'Work, Play and Technology' (*The Philosophical Forum* 10, nos. 2–4 (Winter–Summer, 1978–1979), 321–340), draws upon Hegel and Heidegger to argue that managers play rather than work in a technological society. Adina Schwartz prefers worker autonomy over an unrestricted quest for productivity in 'Meaningful Work', *Ethics* **92** (1982). Edmund Byrne views the displacement of all workers as the ultimate goal implicit in managerial workforce policy in 'Utopia without Work? Myth, Machines and Public Policy', *Research in Philosophy and Technology* vol. 8, ed. P. T. Durbin (Greenwich, CT: JAI Press, 1985, pp. 133–148). Jon Elster analyzes Marxist theories, among others, in *Explaining Technical Change: A Case Study in the Philosophy of Science* (Cambridge: Cambridge University Press, 1983).

More generally, labor saving technology may encourage policies that regulate access to employment, e.g., those with regard to welfare, immigration, and equal employment opportunity. Whence the relevance here of studies of employment-related issues by ethicists and social and political philosophers. Most of these tend to support the *status quo*, in part because the authors seldom address the underlying issues about work. An exception in this regard is James W. Nickel, 'Is There a Human Right to Employment?' in *The Philosophical Forum* **10** (Winter–Summer, 1978–1979). Nickel, however, makes wealth a precondition for any such right. Compare Barbara Dinham and Colin Hines, *Agribusiness in Africa* (Trenton, NJ: Africa World Press, 1984).

Illustrative of the tendency among philosophers to rationalize the *status quo* with regard to employment are two anthologies: Marshall Cohen, Thomas Nagel, and Thomas Scanlon, eds., *Equality and Preferential Treatment* (Princeton, NJ: Princeton University Press, 1976), and Barry Gross, ed., *Reverse Discrimination* (Buffalo, NY: Prometheus, 1977). More balanced than works by Nicholas Capaldi and Alan H. Goldman on this subject is Robert K. Fullinwider, *The Reverse Discrimination Controversy: A Moral and Legal Analysis* (Totowa, NJ: Rowman & Littlefield, 1980).

Serious philosophical consideration of this issue should take into account the data studied in Jonathan H. Turner and Charles E. Starnes, *Inequality: Privilege and Poverty in America* (Santa Monica, CA: Goodyear, 1976), the reflections on equality in *The Concept of Equality*, ed. William T. Blackstone (Minneapolis: Burgess, 1969), the meticulous analysis by Douglas Rae *et al.*, entitled *Equalities* (Cambridge, MA: Harvard University Press, 1981), and the legal survey by Kent Greenawalt, *Discrimination and Reverse Discrimination* (New York: Knopf, 1983).

Frances Fox Piven and Richard A. Cloward elucidate the relationship between welfare policy and work in *Regulating the Poor* (New York: Vintage, 1971). For historical background see Karl de Schweinitz, *England's Road to Social Security, 1349 to 1947* (Philadelphia: University of Pennsylvania Press, 1947); Martha J. Soltow and Susan Gravelle, *Worker Benefits: Industrial Welfare in America, 1900–1935* (Metuchen, NJ: Scarecrow, 1983). Two philosophers have addressed the question of welfare rights in some detail: Nicholas Rescher looks for criteria in *Welfare Rights* (Pittsburgh: University of Pittsburgh Press, 1972), and Carl Wellman proceeds more analytically in *Welfare Rights* (Totowa, NJ: Rowman and Littlefield, 1982). U.S. immigration policies are the subject of two works: John Scanlan and Gil Loescher, *Calculated Kindness* (New York: Free Press, 1986); and *The Border That Joins: Mexican Migrants and U.S. Responsibility*, eds. Peter G. Brown and Henry Shue (Totowa, NJ: Rowman & Allanheld, 1982).

2. THE VALUE OF WORK

As technology makes possible less labor-intensive production, the value of work is inevitably called into question. This questioning has not resulted in any uniform or consistent set of answers. But the types of responses are comparatively few in number. Here they may be reduced to two: (a) the *synchronist* view that technology has changed only the conditions of work (for better or worse) and not the traditional value of work; and (b) the *diachronist* view that technology makes traditional evaluations of work irrelevant and inappropriate. Not falling neatly on either side are (c) certain feminist studies, especially those that deal explicitly with the basic concepts of patriarchy or reproduction.

A. *Synchronist Evaluations of Work*

Characteristic of the synchronist evaluation of work is the body of literature that stresses the value of craftsmanship even in a technologically transformed environment. This nostalgia for craftsmanship is seen today in such works as D. M. Dooling, ed., *A Way of Working* (Garden City, NY: Doubleday, 1979), Tracy Kidder, *The Soul of a New Machine* (Boston: Little, Brown, 1981; New York: Avon, 1982), and a century ago in the writings of British men of letters William Morris and John Ruskin.

Robert M. Pirsig's philosophical novel, *Zen and the Art of Motorcycle Maintenance* (New York: William Morrow, 1974), popularizes the view of all sorts of industrial therapists that proper attitude is the key to satisfaction in a technology-intensive environment. Norman Mailer even links this satisfaction to masculinity in his *Of a Fire on the Moon* (New York: NAL/Signet, 1971). This view is reminiscent of nineteenth century American writers, as studied by Leo Marx in *The Machine in the Garden* (Oxford: Oxford University Press,1964). The failure of these writers to acknowledge that the human factor can be overwhelmed by technology is counterbalanced by Jacques Ellul's well-known warnings and by E. F. Schumacher's reasoned insistence that the human need for work should continue to be met by limiting the scale of technology to what is by that very test "appropriate," e.g., in *Small Is Beautiful* (New York: Harper & Row, 1973) and *Good Work* (New York: Harper & Row, 1979).

Implicit in these defenses of the value of human-scale work is a normative claim made explicit under the heading of the so-called **work ethic**. Current debate about the work ethic centers around comparisons between the commitment of workers today and that of workers in the past with regard to the value of work. Some writers think we are not as dedicated to our work as people used to be. David Charrington chastises Americans on this score, in *The Work Ethic: Working Values and Values that Work* (New York: ANACOM, 1980), and expresses the hope that proper education can remedy the situation. Similar concerns have been expressed with regard to workers in other countries: Lee Smith,'Cracks in the Japanese Work Ethic', *Fortune* **109** (May 14, 1984), pp. 162–168; and Susan Field, 'Egypt's Worst Enemy May be Euphoria', *Euromoney* (UK) (April 1979), pp. 77–82. Michael Rose, on the other hand, insists that lack of dedication to the work ethic among contemporary workers in no way distinguishes them from their forebears: *Reworking the Work Ethic* (London: Batsford Academic, 1985). See also E. Jordan Blakely, *Work Ethic: Pride, or Mental Illness*, ed. M. Sarah Ross (Flint, MI: Jordan Blakely, 1985); and *Work Ethic: An Analytical View, 1983* (Madison, WI: Industrial Research Association, 1983).

The work ethic itself seems to have been an invention of intellectuals, who are generally supportive of its redeeming social features. Max Weber's *The Protestant Ethic and the Spirit of Capitalism* (trans. Talcott Parsons; New York: Scribner's, 1958; German original, 1904–1905) found motivation for entrepreneurial capitalism in the Puritan version of Calvinism: the theological notion of predestination is tied to the earthly goal of financial success. R. H. Tawney's response to Weber, *The Acquisitive Society* (New York, 1920), is the source of E. F. Schumacher's defense of workers' rights as against entrepreneurial irresponsibility. Emile Durkheim's *The Division of Labor in Society* (5 editions, 1893–1926; trans. George Simpson; New York: Macmillan Free Press, 1933) is an early attempt to base a system of ethics on an analysis of work relationships. In particular, Durkheim claims that the division of labor requires us to excel at our own specialized work both as individuals and as members of a professional group.

Durkheim's view is similar to that of Confucianism, but the reason for this cross-cultural similarity is not so easy to identify. A number of recent studies have shown how a work ethic may be founded upon the cultural heritage of a non-Western culture. See, for example, Francis L. K. Hsu, 'Filial Piety in Japan and China: Borrowing, Variation and Significance', *Journal of Comparative Family Studies* (Spring, 1971): 67–74; J. Elder, 'The Gandhian Ethic of Work in India', in *Religious Ferment in Asia*, ed. Robert J. Miller (Lawrence, KS: University of Kansas Press, 1974); and Winston L. King, 'A Christian and a Japanese-Buddhist Work-Ethic Compared', *Religion* 11 (July, 1981): 207–226. For a more general view of the culture-specific role of a work ethic, see Erik von Kuehnelt-Leddihn, 'La Morale du Travail: Un Problème Mondial', *Cahiers de Sociologie Economique* 2: 2 (December, 1971): 215–227.

B. *Diachronist Evaluations of Work*

Missing from synchronist evaluations of work is a willingness to acknowledge that technology has significantly and perhaps irrevocably transformed the conditions of work out of which traditions such as the work ethic emerged. Diachronist evaluations start with this assumption and attempt to formulate alternative norms.

A typical pro-management response is that of Michael Maccoby, *The Leader* (New York: Simon & Schuster, 1981): the "craft ethic," a secularized version of the work ethic, ceased to be important in America when entrepreneurs systematized first unskilled labor and then technology to move beyond dependence upon craftsmanship to leadership. Lacking to this view is the historical perspective of Carl Bridenbaugh's *The Colonial Craftsman* (New York and London: University of Chicago Press, 1961; original, 1950). But neither is it as austere as Nathan D. Grundstein's attempt to justify an authoritarian approach "teleologically": *The Managerial Kant: The Kant Critiques and the Managerial Order* (Cleveland: Case Western Reserve University Press, 1981).

More pro-labor are the responses of historians and social scientists who take into account the **background issue of who controls work**. Attempts to impose a work ethic on workers in the United States have been studied, e.g., by Daniel T. Rogers, *The Work Ethic in Industrial America, 1850–1920* (Chicago: University of Chicago Press, 1978), and by James B. Gilbert, *Work without Salvation: America's Intellectuals and Industrial Alienation, 1880–1910* (Baltimore and London: Johns Hopkins University Press, 1977). A more global perspective is provided by Reinhard Bendix's 1956 study, *Work and Authority in Industry* (republished Berkeley: University of California Press, 1974). Still unsurpassed for its analysis of the relationship between ownership and control of a workplace is Adolf A. Berle and Gardiner C. Means, *The Modern Corporation and Private Property* (rev. ed.; New York: Harcourt, Brace & World, 1968; original, 1932), in which a strong case is made for granting property rights to the community in which a business is located. Comparable views are expressed by more recent analysts of corporate power, but philosophical works tend to repeat the arguments of an earlier era. See, for example, Lawrence G. Becker, *Property Rights: Philosophic Foundations* (London: Routledge & Kegan Paul, 1980).

The foregoing studies of work control presuppose an ongoing workplace. The instability of employment in the face of new technology and other related factors is also being taken into account, at least indirectly, in **philosophical considerations of corporation and employment law**. Much of this material is being incorporated into business ethics texts.

But not all of these are equally attentive to the interests of workers. Exceptions include Patricia Werhane, *Persons, Rights and Corporations* (Englewood Cliffs, NJ: Prentice-Hall, 1985); Thomas Donaldson, *Corporations and Morality* (Englewood Cliffs, NJ: Prentice-Hall, 1982); and Milton Snoeyenbos, Robert Almeder, and James Humber, eds., *Business Ethics* (Buffalo, NY: Prometheus, 1983). Almost half of the latter book deals with employee rights and obligations and hiring and discharge questions. Werhane offers a balanced treatment of the rights and obligations of (a) a corporation and (b) its employees; but she writes as though recognition of a right is equivalent to its enforcement. Donaldson focuses on questions of corporate responsibility, an increasingly debated topic among philosophers; e.g., see the contributions to *Corrigible Corporations and Unruly Laws*, ed. Brent Fisse and Peter A. French (San Antonio, TX: Trinity University Press, 1985).

This being a rapidly changing area of law, law texts need to be regularly supplemented. Useful, however, as an introduction to American law on the subject is Alan E. Westin and Stephan Salisbury, eds., *Individual Rights in the Corporation: A Reader on Employee Rights* (New York: Pantheon, 1980); and to British law: Paul O'Higgins, *Workers' Rights* (London: Arrow, 1976); and Jeremy McMullen, *Rights at Work* (2nd impression with supplement; London: Pluto, 1979). A more technical treatment of American employment law for which updates would be available is Lex K. Larson and Philip Borowsky, *Unjust Dismissal* (Albany, NY: Matthew Bender, 1985).

The foregoing approaches disregard situations in which the workers are unionized. James B. Atleson's *Values and Assumptions in American Labor Law* (Amherst, MA: University of Massachusetts Press, 1983) assumes the existence of unions and shows that common-law endorsement of management rights in the United States is being perpetuated in judicial interpretations of statutory law that on its face clearly grants certain rights to employee unions. Even more intentionally philosophical in their analysis of labor law are two studies of unions as groups: Frank Tannenbaum, *A Philosophy of Labor* (New York: Knopf, 1951), and the doctoral dissertation of John Herman Randall, Jr., *The Problems of Group Responsibility to Society* (reprint; New York: Arno Press, 1969; original, 1922).

More basic still is the claim that the introduction of technology into the workplace renders obsolete any encomium of a work ethic for the simple reason that there is no longer enough work to go around. This concern is typically answered by asserting that new technology creates as many jobs as it eliminates. This comforting view is challenged by such studies as Barry Bluestone and Bennett Harrison, *The Deindustrialization of America* (New York: Basic Books, 1982), and Ian Benson and John Lloyd, *New Technology and Industrial Change* (London: Kogan Paul, 1983); and the debate goes on. Of philosophical interest in this debate is its impact on the work ethic and its various corollaries, e.g., with regard to welfare.

A useful guide to this issue is David Macarov, *Work and Welfare: The Unholy Alliance* (London/Beverly Hills: Sage, 1980). Macarov notes how we have tied welfare (meaning our personal well-being in society) to work on the basis of four specific attitudes, which he identifies as: (1) believing in the myth of needed work; (2) acquiescing in "the job scramble"; (3) viewing work as normalcy; and (4) subsuming work under morality. He calls for clear distinctions among the different meanings of work as a prerequisite to revising our attitudes about work in preparation for increased leisure leading to "an (almost) workless world."

Attention to this very possibility is the hallmark of Hannah Arendt's erudite reconsideration of ancient and mediaeval thought about work: *The Human Condition* (Garden City, NY: Doubleday, 1959). Arendt traces modern attitudes about work to the views of the ancient Greeks, for whom work, unlike labor, results in a product. Automation renders that distinction obsolete but reestablishes the primacy of contemplation over action. This brilliant but somewhat forced explanation of our contemporary crisis needs to be modified in light of such works as Jacques Le Goff's *Time, Work & Culture in the Middle Ages* (Chicago: University of Chicago Press, 1980).

C. *Feminist Reevaluations of Work*

Feminist theory as a whole might be viewed as a reinterpretation of received doctrines regarding the appropriate allocation and evaluation of work. Explicit feminist consideration of work typically involves in some way a comparative evaluation of the work process. What is compared, however, is not the value of work past and present but the value of work commonly done by males and that commonly done by females. In this literature, accordingly, such notions as patriarchy and reproduction serve as lodestones for reinterpreting arguably male supremacist theories.

Alison Jaggar in *Feminist Politics and Human Nature* (Totowa, NJ: Rowman & Allanheld; Brighton: Harvester, 1983) discusses the different ways in which radical, socialist, and liberal feminists, respectively, account for women's subordination in terms of reproduction. In *Capitalist Patriarchy and the Case for Socialist Feminism*, ed. Zillah R. Eisenstein (New York and London: Monthly Review Press, 1979), socialist feminists trace work relationships in the home to capitalist patriarchy. Zillah R. Eisenstein, in *The Radical Future of Liberal Feminism* (New York and London: Longman, 1981), charges liberal feminists with accepting patriarchy and thereby committing the working mother to a "double day." Diverse arguments in support of androgynous co-sharing of sex roles are presented by Mary O'Brien, *The Politics of Reproduction* (Boston/London: Routledge & Kegan Paul, 1981), Nancy Chodorow, *The Reproduction of Mothering: Psychoanalysis and the Sociology of Gender* (Berkeley: University of California Press, 1978), and Dorothy Dinnerstein, *The Mermaid and the Minotaur: Sexual Arrangements and Human Malaise* (New York: Harper & Row, 1976).

These feminist studies should be viewed on a broader canvas that includes the ideology of creativity and innovation, in which many scholars, notably Freud, have seen male supremacist implications. Carolyn Merchant uncovers some pseudo-scientific underpinnings in *The Death of Nature* (New York: Harper & Row, 1979). Mircea Eliade explores the sexual symbolism of ancient mining and metallurgy in *The Forge and the Crucible* (trans. Stephen Corrin; London: Rider, 1962).

3. WORK AND TECHNOLOGY IN THE FUTURE

Work, as organized and divided into jobs, is the principal source of livelihood for many people; and as such it has long been encouraged by every institutional form of persuasion, generally in terms of the work ethic. But technology is transforming work into a scarce and possibly non-renewable resource. This, briefly, is the dilemma facing anyone who wants to prescribe an appropriate relationship between work and technology in the future.

Larry Hirschhorn solves the problem by denial. He contends in *Beyond Mechanization: Work and Technology in the Post-Industrial Age* (Cambridge, MA: MIT Press, 1984) that much of the talk about workers being phased out by automation is greatly exaggerated, because the new electronic technologies which are at its base actually call for more rather than less human involvement. However, workers will need to be better educated to deal with the complex demands that machines make on humans. Comparable optimism is expressed by Theodore Roszak in *The Cult of Information* (New York: Pantheon, 1986), Alvin Toffler in *The New Wave* (New York: William Morrow, 1980), and John Naisbitt in *Megatrends* (New York: Warner, 1982). More sensitive to structural complexities are the contributions to *Futures for Work*, ed. Geert Hofstede (The Hague: Martinus Nijhoff, 1979), and *The World of Work: Careers and the Future*, ed. Howard F. Didsbury, Jr. (Bethesda, MD: World Future Society, 1983).

Generally indifferent to the methodological pitfalls of prediction, philosophers who have addressed this subject tend to assume that full-time meaningful work will not be available to all who need or want it in the future.

Bernard Gendron in *Technology and the Human Condition* (New York: St. Martin's Press, 1977) and David Schweickart in *Capitalism or Worker Control?* (New York: Praeger, 1980) both argue that a version of worker control is preferable to what they view as the unacceptable impact of capitalist arrangements on the workplace.

Herbert Marcuse in *One-Dimensional Man* (Boston: Beacon, 1964), recognizes the short-range concerns of workers, but insists that long-term opposition to automation stands in the way of eventual attainment of a liberating utopia based on technology. This latter, however, cannot occur under capitalism: see 'The End of Utopia', one of *Five Lectures* by Marcuse (trans. J. J. Shapiro and S. M. Weber; Boston: Beacon, 1970).

André Gorz says technology will lead to "liberation from work"; this, however, can be achieved only within a social environment which does not yet exist (at least not generally): *Paths to Paradise: On the Liberation from Work* (London and Sydney: Pluto Press, 1985). In aid of this movement *ad astra*, Frithjof Bergmann recommends organizing the unemployed politically: see 'The Future of Work', *Praxis International* 3 (October, 1983): 308–323.

Jonathan Glover (see Chap. 9, 'Work', in his *What Sort of People Should There Be?* (New York: Penguin, 1984) envisions a society in which machines do the unpleasant work, people share other jobs, and sufficient income is made available to all. To assure that people will have intrinsically satisfying work, society may need to give preference to humans over machines even if the latter could replace the former. This view echoes that put forward a century earlier by William Morris in his novel *News from Nowhere* (London, 1891).

David Braybrooke, in 'Work: A Cultural Ideal Ever More in Jeopardy' (*Midwest Studies in Philosophy*, vol. 7, eds. Peter A. French *et al.*; Minneapolis: University of Minnesota Press, 1982), shares Glover's concern about retaining meaningful work, and suggests that utilitarianism may have to provide the basis for a future "ethics of welfare."

Edmund Byrne, in 'Displaced Workers: America's Unpaid Debt', *Journal of Business Ethics* **4** (1984): 31–41, argues for a national commitment to solving the problem of displaced workers because business alone cannot provide enough fulltime paying jobs in the future. Comparable concerns are expressed in Francis X. Quinn, ed., *The Ethical Aftermath of Automation* (Westminster, MD: Newman, 1962).

More generally, Australian Barry Jones's *Sleepers, Wake! Technology and the Future of Work* (Melbourne: Oxford University Press, 1982) is perhaps the best recent treatment of the futurist issue, partly because it includes surveys of the impact and the value issues. Although drawing heavily on Australian data, Jones knowledgeably covers each of the three principal areas, concludes with a 22-point program for the future, and emphasizes the importance of consciousness-raising. In short, probably the one best overall introduction to the issues.

On none of these issues, of course, has the last word been said; so there is ample opportunity to add additional philosophical perspective to the complex interface between work and technology. In particular, this bibliographical review covers neither workplace health and safety, philosophical aspects of which have been studied, for example, by Mary Gibson, Nicholas Rescher, and Kristin Shrader-Frechette, nor the problem of exploitation in developing countries. Nor does it represent a systematic search of the literature in languages other than English. With regard to areas that have been covered, the author welcomes reminders of significant omissions for which he is undoubtedly but not willfully responsible.

Indiana University/Purdue University at Indianapolis

PHILOSOPHY AND TECHNOLOGY

Series Editor: PAUL T. DURBIN

1. Paul T. Durbin and Friedrich Rapp (eds.), *Philosophy and Technology,* 1983.
2. Carl Mitcham and Alois Huning (eds.), *Philosophy and Technology II: Information Technology and Computers in Theory and Practice,* 1986.
3. Paul T. Durbin (ed.), *Technology and Responsibility,* 1987.
4. Paul T. Durbin (ed.), *Technology and Contemporary Life,* 1988.

NAME INDEX

[*Note*: Except for the Barzel and Levinson articles, notes and references are not cited here.]

Abendroth, Wolfgang 299
Abrams, P. 59
Adams, Henry 9
Adler, Margot 172–174
Agassi, Judith Buber 299
al-Hibri, Azizah 188
Almeder, Robert 309
Althusser, Louis 19, 61, 190, 298
Altman, Irwin 195
Andreae, Johann 70
Angang, S. O. 108
Anscombe, Elizabeth 229
Aquinas, Thomas 68
Arendt, Hannah 161–162, 175, 217, 219, 220–223, 229, 231, 232, 310
Aristotle 154, 243, 295
Armstrong, D. M. 226, 229
Atleson, James B. 309
Axelos, Kostas 298

Babbage, Charles 77, 297
Bacon, Francis 280–281
Bagrit, Leon 301
Baran, Paul A. 29
Barbour, I. 60
Barnett, George E. 302
Barry, Brian 37, 38
Bartolke, K. 59
Becker, Lawrence G. 308
Begin, Menachim 37
Bell, Daniel 301
Bellah, Robert 38
Bellers, John 71
Bendix, Reinhard 308
Benson, Ian 310
Bergmann, Frithjof 312
Bergmann, T. 59
Berle, Adolf A. 308
Bernard, Cheryl 111
Best, Michael 20
Blackstone, William T. 305

Blakely, E. Jordan 307
Bloom, Leopold 166
Bluestone, Barry 310
Bochenski, J. M. 237
Bolger, Thomas E. 187
Bookchin, Murray 117–118, 124–125
Borgmann, Albert 1–11, 101
Borowsky, Philip 309
Bossel, H. 59
Bourbaki, N. 241
Boyle, Godfrey 123
Braverman, Harry 18, 29, 73–74, 80, 298, 303
Braybrooke, David 313
Bridenbaugh, Carl 308
Brooks, Harvey 88, 92–95, 100, 116
Brown, Peter G. 305
Browning, Douglas 164
Buder, Stanley 297
Bugliarello, George 59
Burns, P. 201
Burton, Robert 70
Bush, Vannevar 77
Bussiek, H. 62
Byrne, Edmund 304, 313

Cabet, Etienne 61
Calvin, John 67, 70
Campanella, Tommaso 61
Capaldi, Nicholas 304
Carnap, Rudolf 223
Carney, J. D. 244
Carpenter, Stanley R. 29, 37–39, 42
Caws, Peter 59
Chamot, Dennis 302
Chapuis, Alfred 300
Charrington, David 306
Chase, Stuart 302
Cherns, A. 59, 62
Cherry, Colin 190
Chodorow, Nancy 311

315

Clark, Robin 110
Clarke, Arthur C. 177
Claude, Georges 247
Clawson, Dan 303
Cloward, Richard A. 305
Coates, Joseph 164
Cock, P. 59
Cohen, Marshall 304
Cohen, N. 60
Collins, Joseph 114, 124
Comte, Auguste 196, 221
Connolly, William 19–20
Copi, I. M. 245
Courtney, Frederic 144
Cowan, Ruth Schwartz 303
Crowley, Ambrose 71
Csikszentmihalyi, Mihaly 163
Cybele 172

Dahl, Robert 125
Davidson, Donald 228–230
Davis, J. C. 297
de Butts, John D. 187
Dede, Christopher 188
De Forest, Paul H. 112, 114–115
de Hahn, David 188
de la Mettrie, J. O. 300
Descartes, René 87, 287
de Schweinitz, Karl 305
Dessauer, Friedrich 59, 280
Dewey, John 163, 172
de Wilde, Tom 110
Dickson, David 114
Didsbury, Howard Jr. 188, 312
Diebold, John 301
Dinham, Barbara 304
Dinnerstein, Dorothy 311
Diwan, Romesh K. 110
Domhoff, William 29
Donaldson, Thomas 309
Doner, Dean B. 59
Dooling, D. M. 306
Droz, Edmund 300
Ducasse, C. J. 228–229
Dummett, Michael 222
Dunlap, Elliott 302
Durbin, Paul T. 189, 304

Durkheim, Emile 307
Dworkin, Ronald 18, 38
Dymmel, Michael D. 302

Eberlein 70
Edwards, Richard 299
Eisenstein, Elizabeth 187
Eisenstein, Zillah R. 311
Elder, J. 307
Eliade, Mircea 311
Elliott, David 124
Ellul, Jacques 162, 174, 190, 296, 306
Elsner, Henry 300
Elster, Jon 304
Engelberger, Joseph P. 300
Engels, Friedrich 19, 61, 80, 296, 299
Evans, John 302
Evans, Lawrence B. 75
Ewen, Stuart 29

Fetscher, I. 62
Field, Susan 306
Fischoff, B. 262
Fisse, Brent 309
Florman, Samuel 190–191
Ford, Henry 16, 93, 111
Fourier, Charles 61
Francoeur, Robert 189
French, Peter A. 309, 313
Freud, Sigmund 81, 311
Friedmann, Georges 302
Fromm, Erich 82, 298
Fuller, Buckminster R. 62, 190–191
Fullinwider, Robert K. 304

Galbraith, John Kenneth 7
Galileo 172, 241
Gassman, H. P. 209
Geertz, Clifford 296
Gehlen, Arnold 282
Gendron, Bernard 312
Genovese, Eugene 19
George, Henry 302
Georgescu-Roegen, Nicholas 98–99
Geyer, R. Felix 298
Gibson, Dale 205
Gibson, Mary 313

Gilbert, James B. 308
Gill, Colin 301
Gillett, Mr. 204–205
Gilsen, Michael 189
Glover, Jonathan 312–313
Godelier, Maurice 19
Goldman, Alan H. 304
Goodlad, John 188
Goodman, Paul 62
Gordon, David M. 299
Gordon, Frederick M. 299
Gorz, Andre 70, 81, 299, 312
Graham, J. 268–269
Gravelle, Susan 305
Greenawalt, Kent 305
Greenbaum, Joan M. 302
Gross, Barry 304
Grundstein, Nathan D. 308
Guericke 241
Gutman, Herbert 19

Habakkuk, H. J. 301
Habermas, Jurgen 18–19, 29, 190
Habicht, H. 60
Hall, Edward 163
Harper 123
Harrison, Annette 302
Harrison, Bennett 310
Harvey, David 17, 19
Hausman, Daniel M. 298
Havelock, Eric 187
Hegel 154, 304
Heidegger, Martin 3–4, 14–16, 145–157,
    219, 220, 304
Heilbroner 9, 10, 60
Heimerdinger, M. 188
Heinz, Walter R. 298
Heisenberg, Werner 218–219, 220
Hempel 223–226
Henry VIII 69
Hesse, Mary 245
Hickman, Larry 188
Hill, Stephen 301
Hilier, Egmont 300
Hiltz, Starr Roxanne 188–190
Hindess, Barry 297
Hines, Colin 304

Hirschhorn, Larry 311
Hirst, Paul Q. 297
Hitler, Adolf 37
Hobbes, Thomas 96
Hofstede, Geert 312
Hogarth, S. H. 188
Hölderlin, Friedrich 151
Horton, John 298
Howard, Robert 302
Hsu, Francis L. K. 307
Huelphinil 110
Humber, James 309
Hume 227–229, 244
Huning, Alois 300
Hunt, H. Allan 301
Hunt, Timothy L. 301
Huygens, Christiaan 238
Huxley, Aldous 196

Ihde, Don 59
Illich, Ivan 95
Innis, Harold A. 187
Ireland, Norman J. 298
Israel, Joachim 299

Jackson, Sarah 114
Jacoby, Sanford M. 296
Jaggar, Alison 311
Jaspers, Karl 190
Jequier, Nicholas 109, 123–124
Jesus Christ 67, 170
Jonas, Hans 279
Jones, Barry 313

Kahler, Erich 60
Kamerlingh Onnes, Heike 247
Kant, Immanuel 35, 87, 89, 144, 244,
    287
Kanter, R. 60
Kepler, Johannes 172
Khaddafi, Moammar 59
Khalilzad, Zalmay 111
Khomeini, Ayatollah Ruhollah 59
Kidder, Tracy 306
King, Winston L. 307
Kohn, Leopold 112
Kuhn, Thomas S. 90

Kuic, Vukan 297
Kurtz, Paul 189

Langhaar, H. L. 245
Lappé, Frances Moore 114, 124
Larson, Lex K. 309
Law, Peter J. 298
Lawrence, T. E. 26
LeCarré, John 25
Lee, Alfred 188–189
Le Goff, Jacques 310
Leibniz, G. W. 77, 299
Leiss, William 29, 125
Lenin, V. I. 80
Lenk, Hans 287
Lévi-Strauss, Claude 23
Levidow, Les 303
Lichtenstein, S. 262
Liegle, L. 59
Linde, G. 247
Livingston, David 24
Livingston, Dennis 110
Lloyd, John 310
Locke, John 87, 96, 281
Loescher, Gil 305
Lofthouse, Paul R. 119
Lovins, Amory 92, 94, 100, 113
Ludz, Peter Christian 298
Lukacs, Georg 20, 296
Lull, Ramon 299
Lum, Man-Kong 189
Luther, Martin 69

Macarov, David 310
Maccoby, Michael 308
MacCormac, Earl 283
MacIntyre, Alasdair 4, 40
Macmillan, Robert Hugh 300
Mailer, Norman 306
Mandelbaum, Maurice 225
Manuel, Frank E. 60, 297
Manuel, Fritzie P. 60, 297
Marcson, Simon 299
Marcuse, Herbert 17, 29, 32, 36–37,
    80–81, 190, 312
Margolis, Joseph 59, 91
Markovic, Mihailo 62, 298

Marx, Karl 17, 19, 36, 52, 61, 67, 70,
    73–74, 80–81, 129, 220, 281, 296–298,
    301, 303
Marx, Leo 306
Maxwell, James Clerk 238, 244
McEnnis, John T. 302
McCulloch, A. 59
McLuhan, Marshall 163, 187–188
McMullen, Jeremy 309
McNeill, William 190
McRobie, George 112–115, 122
Mead, G. H. 163
Means, Gardiner, C. 308
Meek, Ronald L. 297
Melville, K. 60
Menzies, Heather 303
Merchant, Carolyn 311
Meyrowitz, Joshua 188
Mill, John Stuart 40, 221–223
Miller, Robert J. 307
Mitcham, Carl 144, 300
Mondale, Walter 31
Monter, E. William 172
Montgomery, David 299
More, Thomas 61
Moreno, Manuel 298
Morris, William 82, 306, 312
Mumford, Lewis 188, 190, 299
Murphy, Kevin 245, 302

Nadworny, Milton J. 303
Nagel, Ernest 244–245
Nagel, Thomas 304
Naisbitt, John 312
Ndongho, W. W. 108
Newcomen, Thomas 238, 246
Newton, Isaac 238
Nickel, James W. 304
Nidal, Abu 26
Nietzsche, Friedrich 291
Nisbet, Robert 97
Noble, David 19, 74, 123, 303

O'Brien, Mary 311
O'Connor, James 19
Offe, Claus 301
O'Higgins, Paul 309

Okrent, Mark 304
Osborne, P. H. 201
Owen, Robert 61

Pacey, Arnold 120–121
Pagano, Ugo 298
Pascal, Blaise 77, 299
Pawlowski, J. 245
Peirce, C. S. 163, 299
Papin, Denis 238
Perrin, Noel 190
Philby, Kim 25
Piercy, Marge 82
Pirsig, Robert M. 306
Piven, Frances Fox 305
Plato 147, 172
Platt, J. R. 296
Plattes, Gabriel 71
Polanyi, Karl 97
Popper, Karl 220–227, 229, 231, 233
Portes, Alejandro 299
Postman, Neil 188, 191
Proudhon, Pierre Joseph 61, 70
Pythagoras 243

Quine, W. V. 90, 225
Quinlan, Karen Ann 132
Quinn, Francis X. 313

Rae, Douglas 305
Randall, John Herman, Jr. 309
Rawls, John 18, 22, 26, 38, 40–41, 265
Reagan, Ronald 30–31, 87
Reich, Michael 299
Rescher, Nicholas 266, 270, 305, 313
Richard II 69
Riesman, David 191
Rigby, A. 60
Roberts, Mr. 72
Rochberg-Halton, Eugene 163
Rodman, George 189
Roemer, John 19
Rogers, Daniel T. 308
Rose, Michael 306
Ross, M. Sarah 307
Rosser-Owen, Dawud G. 107, 117, 126
Roszak, Theodore 312

Rothschild, Joan 303
Ruether, Rosemary 60
Rule, James B. 200
Ruskin, John 82, 306
Russell, Bertrand 89, 224
Rüstow, A. 286
Ryan, E. F. 201
Rybcynski, Witold 107, 110–111, 116

Saint Benedict 67–68
Saint Paul 68
Saint Francis of Assisi 68
Salisbury, Stephan 309
Sandel, Michael 37–38
Sardar, Ziauddin 107, 117, 126
Sattler, K. 62
Savery, Thomas 238, 246
Scanlan, John 305
Scanlon, Thomas 304
Schacht, Richard 298
Scheer, R. K. 244
Schell, Jonathan 25
Schumacher, E. F. 5, 8, 62, 82, 92–95,
    97–98, 100, 108, 111–112, 114, 116,
    120, 306–307
Schwartz, Adina 304
Schweickart, David 312
Schweitzer, David 298
Seligman, Ben B. 302
Shaiken, Harley 302
Shaikh, Rashid 112, 123–124
Shakow, D. 268–269
Shar, Shimon 60
Sheehan, George 23
Shepard, Jon M. 299
Shrader-Frechette, Kristin 313
Shue, Henry 305
Shuttleworth, John 100
Sieferle, R. P. 283, 285
Simon, Herbert A. 301
Simon, Yves R. 297
Simonds, John Cameron 302
Simpson, George 307
Skillen, Anthony 304
Sklar, Robert 188
Slovic, Paul 262
Smart, J. J. C. 226

Smith, Adam 9, 71, 99, 101, 259
Smith, Lee 306
Snoeyenbos, Milton 309
Sollow, Benjamin 303
Soltow, Martha J. 305
Stalin, Joseph 37, 59
Stanley, Manfred 29, 33–34, 37–39, 41–42
Starnes, Charles E. 305
Starr, Chauncey 259, 262, 271–273
Stewart, Frances 109, 114
Strasser, Susan 303
Sweezy, Paul M. 18, 29
Szucs, E. 245

Tannenbaum, Frank 309
Taviss, Irene 300
Tawney, R. H. 307
Taylor, Charles 96, 99–101
Taylor, Frederick Winslow 74, 297
Taylor, G. 62
Teilhard de Chardin, Pierre 190
Terborgh, George 301
Theobald, Robert 27, 80
Thomas, Sari 190
Thompson, E. P. 19
Thury, Eva 189
Toffler, Alvin 312
Tonnies, Ferdinand 52, 59
Torquemada, Tomas 37
Torrance, John 298
Turing, A. M. 299
Turner, Jonathan H. 305
Turoff, Murray 188

Updike, John 163–166, 169, 173–174
Ure, Andrew 70, 72, 74–75, 78

van Brakel, J. 110
Veblen, Thorstein 95, 282, 297
Vilmar, Fritz 62
Viscusi, W. K. 260–262, 264
Voltaire, Francois 65
von Gizycki, H. 60
von Kuehnelt-Leddihn, Erik 307

Walton, John 299
Wartofsky, Marx 59, 190
Watt, James 238, 246
Weber, Max 21, 66–67, 70, 279, 296, 307
Weinberg, Alvin 111
Welch, Lawrence 188
Wellman, Carl 305
Werhane, Patricia 309
Westin, Alan 194–195, 309
Whipple 262, 271, 273
White, Lynn, Jr. 1, 9
Wiener, Norbert 302
Wigner, Eugene 111
Wilkinson, John 60
Wilson, Jackson 61
Winner, Langdon 2, 10, 87–88, 91, 94, 101–111, 116–117
Wisser, Richard 145
Wolff, Robert Paul 19, 298
Wood, Stephen 304
Woolfson, Charles 296
Wordsworth, William 82

Young, Bob 303
Young, John Richard 41
Young, Thomas 238
Zimbalist, Andrew 303
Zimmerli, W. Ch. 280, 282, 285, 288
Zablocki, B. 60